Lung Disease in Pregnancy 101:
A Patient Primer

Arunabh Talwar, M.D., F.C.C.P.

Co-Editors:

Syed Mohammed Rizvi, M.S.

Sameer Verma, M.D.

With Contributions From

Ranjit Singh B.S.

Prasanth Kurup B.S.

Shilpa Malik M.M.S.

Hiral Shah B.S.

Shruti Singh M.D.

Kritika Khanna M.D.

This project was made possible by the
Cynthia and Michael Rubinberg
Community Innovation Grant

From the

Katz Institute for Women's Health
Northwell Health

Disclaimer

This book is not intended to replace a physician's advice. While medical professionals are writing this book, there is no substitute for personalized medical care. Every person has individual needs and not every statement in this book will apply to your personal health. You should consult your doctor before making any changes in your healthcare or following any advice you may find in this book.

Table of Contents

Chapter 1:   Body Changes of Pregnancy   Page 9

Chapter 2:   Breathing Basics   Page 13

Chapter 3:   Diseases of the Lung in Pregnancy   Page 15

Chapter 4:   Diagnosing your Disease   Page 33

Chapter 5:   Pulmonary Medications and Pregnancy Page 37

Chapter 6:   Exercise During Pregnancy   Page 45

Chapter 7:   Health Maintenance   Page 57

Chapter 8:   Cardiac Disease in Pregnancy   Page 65

Chapter 9:   Pregnancy & Other Health Problems   Page 71

Chapter 10:   Pregnancy and Tobacco   Page 81

Chapter 11:   Diet and Nutrition   Page 89

Chapter 12:   Traveling During Pregnancy   Page 97

Chapter 13:   Pregnancy and Mental Health   Page 101

Appendix   Page 105

Suggested Readings   Page 109

1 | Body Changes of Pregnancy

"Giving birth and being born brings us into the essence of creation, where the human spirit is courageous and bold and the body, a miracle of wisdom."

Harriette Hartigan

Pregnancy and birth are often regarded as "The Miracle of Life" and in many ways it is. The body undergoes many changes during pregnancy to accommodate the growth and nourishment of your growing child. This chapter will focus on the changes that the different systems in your body will make to ensure a successful pregnancy.

Respiratory System Changes

During pregnancy, the lungs are one of the main organs that are affected. The respiratory rate rises to compensate for increased maternal oxygen consumption, which is caused by the demands of the uterus, the placenta, and the fetus. In general, the physiologic changes of the respiratory system are:

- Increased rate of breathing
- Increased minute ventilation
- Decreased capacity of lungs to expand
- Increased tidal volume
- Shortness of breath

Cardiovascular System Changes

The heart and blood vessels are also heavily affected by pregnancy. During pregnancy, the entire cardiovascular system is readjusted. Blood volume increases greatly and the expanding uterus presses on the large veins. This results in:

- Increased cardiac output
- Increased blood volume
- Elevated resting heart rate
- Decreased blood pressure (second trimester)

Gastrointestinal System Changes

As the uterus enlarges, it rises up and out of the pelvic cavity. This action displaces the stomach, intestines, and other adjacent organs. Progesterone, a hormone released by the uterus at increased rates during pregnancy, causes relaxation of the lower esophageal sphincter. This can lead to some of the gastrointestinal changes listed below:

- Decreased stomach motility/constipation
- Increased reflux
- Heartburn

Endocrine System Changes

The endocrine system of your body, including all the glands that make hormones, goes under significant changes as well. Hormonal changes readjust the entire body system.

The placenta acts as a temporary endocrine gland during pregnancy. It produces large amounts of estrogen and progesterone by the 10th to 12th week of pregnancy. Estrogen and progesterone are hormones that are used to help regulate the health of the female reproductive system. The placenta supports the growth of the fetus and uterus.

- Pregnant women may feel warmer or experience "hot flashes" caused by increased hormonal level and basal metabolic rate.
- The parathyroid gland increases in size slightly to meet the increased requirements for calcium.
- Near the end of term, the posterior pituitary will begin to secrete oxytocin that will serve to initiate labor.
- At birth, the anterior pituitary will begin to secrete prolactin. This stimulates the production of breast milk.

Other Common Changes

Hormonal changes make many women experience changes in both hair and nail texture and growth during pregnancy. In addition, feet and ankles may swell due to extra fluid building up in the body during pregnancy.

Leg cramps can be caused by fatigue from carrying pregnancy weight, compression of the blood vessels in the legs, excess phosphorous, a shortage of calcium or magnesium, and fluctuation of pregnancy hormones. Lastly, there might be a slight increase in body temperature in early pregnancy. The temperature returns to normal at about the 16th week of gestation.

General Advice to Patients Regarding Pregnancy

Prenatal Visits

Prenatal care should begin early in the pregnancy thought process and should be maintained throughout pregnancy. You should keep a tight schedule of regular prenatal visits: 4-28 weeks, every 4 weeks; 28-36 weeks, every 2 weeks; 36 weeks on, weekly.

Diet

The patient should consume a balanced diet (Please see chapter 11 on Nutrition).

Medications

Only medications prescribed or authorized by the obstetric provider should be taken.

Alcohol, Tobacco, and Other Drugs

Abstain from alcohol, tobacco, and all recreational drugs. No safe level of alcohol intake has been established for pregnancy.

Smoking is harmful to you and your baby. To read more about tobacco and pregnancy, please see Chapter 10.

Other recreational drugs —e.g. marijuana, cocaine, and amphetamines— can adversely affect the outcomes of pregnancy and should be avoided at all costs.

Vaccinations

Follow vaccination guidelines as suggested by your physician (See Table in Appendix A).

Noxious Exposures

Chemical or radiation hazards should be avoided, as should excessive heat in hot tubs or saunas. In addition, avoid handling cat feces or cat litter and gloves when gardening to avoid infection with toxoplasmosis.

Rest and Exercise

You are encouraged to obtain adequate rest each day. Regular exercise can be continued at a mild to moderate level with caution.

Birth Classes

Enroll in a childbirth preparation class, often called Lamaze class, with your partner or friends, well before your due date.

2 | Breathing Basics

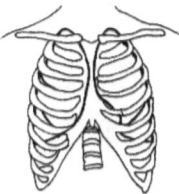

"Breath is the bridge which connects life to consciousness, which unites your body to your thoughts."

Thich Nhat Hanh (b. 1926)

Let's first review the basic mechanisms and anatomical components of breathing. The main functions of the lungs are to help us inhale air, transfer oxygen to our blood, and help remove carbon dioxide from our body.

Anatomical Structures of the Lung

Upper Airways
Includes the nose, mouth, and throat. Air is warmed, filtered, and humidified by these structures as it passes through the upper airways.

Trachea
A tube-like structure allowing air to pass between the throat and the lung, also known as the windpipe.

Bronchi
Two main branches of the trachea leading to the left and right lungs. Bronchi further subdivide in the lungs forming a bronchial tree.

Bronchiole
Smaller branches of bronchi.

Alveoli
Sac-like structures at the end of the bronchioles. They are thin walled allowing for the exchange of oxygen and carbon dioxide between the lungs and the bloodstream.

Interstitium
Area between two alveoli where tiny blood vessels (capillaries) are located.

Mucous Membrane
A thin membrane lining the breathing passages beginning at the nose and ending in the small bronchi covered with mucous producing glands. Mucus is a slimy substance that airways produce to help remove inhaled dust, bacteria, and other small particles.

Diaphragm
A large muscle dividing the abdominal cavity and chest that is the main muscle of breathing.

The primary function of the lung is to transfer oxygen from the air into the body and to remove carbon dioxide. Oxygen acts as the body's fuel, while carbon dioxide is a waste product. Air enters the lungs through the upper airways, eventually reaching the air sacs (alveoli). The alveoli are surrounded by many tiny blood vessels known as capillaries. It is here, at the meeting point between blood vessels and alveoli, that blood is oxygenated and carbon dioxide is released and exhaled.

The normal lung has a mechanism for defending itself against foreign particles. Mucous glands and little hair-like structures, known as cilia are found in the walls lining the bronchial tubes. Whenever a foreign particle enters the lung, it is trapped by a blanket of mucous and wave-like motions of the cilia move it up and away from the lung to the mouth, where it is exhaled by coughing.

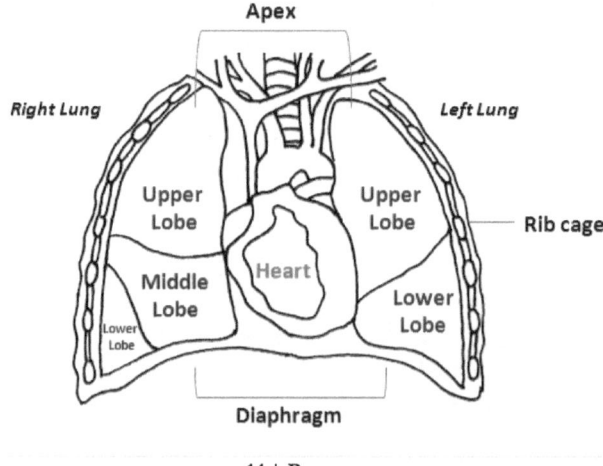

3 | Diseases of the Lung in Pregnancy

"The dreams that we have in pregnancy are tainted with the worries an joys pregnancy and the changing roles of our lives"

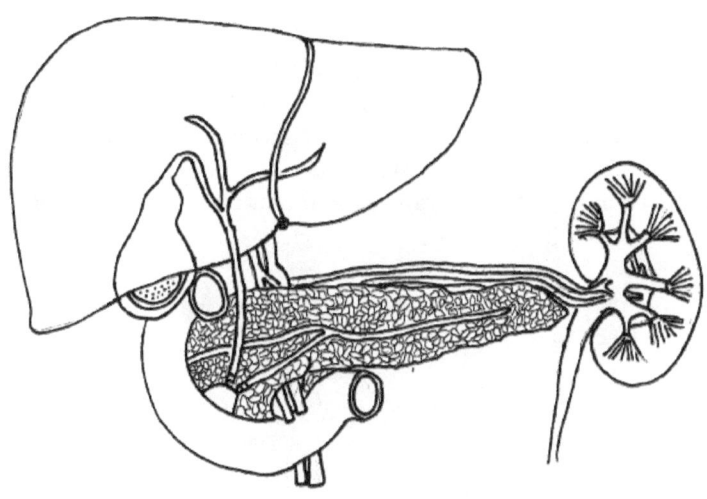

Robin Elise Weiss

There are many existing lung diseases. Let us review a few common lung diseases that result in chronic illness. All chronic lung diseases, if not treated, result in progressive shortness of breath (SOB). Below you will find a few examples of lung disease.

Common Examples of Lung Diseases

Diseases of Upper Airways	Infection of the sinuses, tonsils
Diseases of Lower Airways	Diseases involving trachea, bronchi and bronchioles, Chronic obstructive pulmonary disease (COPD), Asthma, Bronchiectasis (inflammation and dilatation of airways)
Pulmonary Artery	Pulmonary Arterial Hypertension, Pulmonary Embolism.
Pleura	Pleural effusion - fluid buildup in the pleura - a structure around the lung
Alveoli	Pneumonia, Viral Flu
Interstitium	Interstitial lung diseases

Asthma

Asthma is a long term condition affecting airways which may be triggered by allergens such as pollens and/or irritants such as viruses and pollution. It is characterized by episodes of chest tightness, wheezing, cough, and shortness of breath. The symptoms are primarily due to tightening of muscles surrounding the airways, inflammation, and irritation of the airways in the lungs.

Asthma is one of the most common pulmonary disorders during pregnancy, affecting 4 to 8 percent of pregnant woman. The typical symptoms of asthma consist of wheezing, chest tightness, and cough. Of the pregnant women with a history of asthma, one-third will experience improvement of symptoms, one-third will remain stable, and one-third will worsen. These changes occur during the 2nd and 3rd trimester with a peak at 6 months, where you might experience an exacerbation of symptoms. The aim of treating asthma during pregnancy is to prevent respiratory failure.

Pregnancy and Asthma

If you keep your asthma under control, it probably won't cause any problems during your pregnancy.

If you don't control your asthma, you may be at risk for a serious health problem called preeclampsia, which is a combination of high blood pressure and protein in your urine. In addition, your baby may not get enough oxygen and may be at higher risk for health problems such as:

- Premature birth
- Poor growth
- Low birth weight (less than 5½ pounds)

What are the Causes of Asthma?

Incidence of asthma has increased dramatically over the last several decades. While the exact cause of asthma is still unknown, many think the following factors may contribute to the development of asthma:

- Atopy, or an inherited tendency to develop allergies.
- Family history of asthma or allergy.
- Contracting certain viral respiratory infections in early childhood.
- Exposure to some airborne allergens (pollens, smoke, pet dander, etc.) and other allergens like dust mites or indoor molds.
- Hyper-reactive airways (exaggerated airway response to stimuli).

Asthma symptoms wax and wane over time, with treatment focused on prevention, control, and reducing airway reactivity and airway inflammation.

What are the Symptoms of Asthma?

The constriction and inflammation may cause patients to experience some or all of the following classical symptoms of asthma:

- Wheezing
- Chest tightness

- Shortness of breath
- Chronic cough

How is Asthma Diagnosed?

Diagnosing asthma requires 2 specific criteria, one of which is the presence of symptoms compatible with asthma. In addition, an objective measurement of decreased airflow in your lungs is done through a specific measurement, like peak expiratory flow and measurements of patient's lung functions.

What is the Treatment for Asthma?

Treatment for your asthma will primarily involve three main components:

- Avoidance of triggers. The first step is recognizing the triggering factors and minimizing the exposure to such triggers. This is essential to controlling asthma symptoms.
- Treatment with medications. Treatment falls into two categories: *long term controllers* and *rescue inhalers*. The controller medications are used over a period of time to minimize symptoms on a daily basis and reduce the chronic inflammation associated with asthma. Rescue inhalers are bronchodilators drugs that instantly improve the symptoms of wheezing. Rescue medications are those that are used specifically when asthma flares up causing an asthma attack. The controller medications can be in the form of inhalers or oral medicines. Of course, even with daily controller medications, you should always have a rescue inhaler with you to quickly relieve symptoms. In some patients, treatment can be focused on preventing the body's immune response to allergy triggers. Physicians may use antibody injections to prevent allergy-associated asthma.
- Monitoring your peak expiratory flow and asthma symptoms. Also, beyond medications, life style changes can be made to minimize the impact of asthma symptoms.

Safety Profile of Asthma Medication

Name	Safe in Pregnancy?	Safe in Breastfeeding?	Comments
Inhaled beta 2 agonist	Y	Y	
Inhaled long-acting beta 2 agonist (in combination with inhaled corticosteroid)	Y	Y	
Inhaled corticosteroid	Y	Y	Pregnancy: High dose therapy unknown
Oral and intravenous theophylline	Y	Y	Pregnancy: Check levels and aim for lower therapeutic range since protein binding decreases in pregnancy, resulting in increased free drug levels Breastfeeding: <1% excreted into breast milk (Turner et al, 1980)

Oral/intravenous corticosteroids	Y	Y	Risks: cleft palate, pre-eclampsia, preterm labor. But severe asthma confounding variable (Gregerson and Ulrik, 2013)
Chromones	Y/N	Y	Pregnancy: Moderate teratogenic risk cannot be excluded (Tata et al, 2008)
Leukotriene antagonists	Y/N	U	Not to commence during pregnancy
Omalizumab	U	U	Pregnancy: Breastfeeding: No data in humans. 1.5% passed into milk in animals. IgG excreted in milk so likely. Caution is needed

*Y, yes; N, no; U, unknown

Cystic Fibrosis

Cystic fibrosis (CF) is a life-threatening, genetic disease that causes persistent lung infections and progressively limits the ability to breathe. In people with CF, a defective gene causes a thick buildup of mucus in the lungs, pancreas, and other organs. In the lungs, the mucus clogs the airways and traps bacteria leading to infections, extensive lung damage, and, eventually, respiratory failure. In the pancreas, the mucus prevents the release of digestive enzymes that allow the body to break down food and absorb vital nutrients.

The survival of patients with cystic fibrosis has improved dramatically over the years. As more children become adults, issues such as fertility and pregnancy become relevant in the adult cystic fibrosis patients. It is now not uncommon for a woman with cystic fibrosis to complete a successful natural delivery.

How is Cystic Fibrosis treated in pregnancy?

If you have CF and are thinking about getting pregnant, talk to your provider. You need ongoing care throughout pregnancy from your provider and other specialists experienced with Cystic Fibrosis.

Can you find out during pregnancy if your baby has CF or is a CF carrier?

Yes. If you or your partner has CF or is a CF carrier, you can have a prenatal test to find out if your baby has the condition or is a carrier. You can have either of these tests:
- Chorionic villus sampling (also called CVS). This test checks tissue from the placenta for birth defects and genetic conditions. You can take a CVS at 10 to 12 weeks of pregnancy.
- Amniocentesis (also called amnio). This test checks amniotic fluid from the amniotic sac around your baby for birth defects and genetic conditions. You can get this test at 15 to 20 weeks of pregnancy.

Talk to your provider or genetic counselor if you're thinking of having either of these tests.

Bronchiectasis

Bronchiectasis is a condition in which damage to the airways causes them to widen and become flabby and scarred. It is usually the result of an infection and results in an injury to the airways or prevents the airways from clearing mucus. When mucus can't be cleared, it builds up and creates an environment in which bacteria can grow, which leads to recurring serious lung infections.

Each infection causes progressively increasing damage to your airways. Overtime, the airways lose their ability to move air in and out, which can prevent adequate oxygenation of organs. Bronchiectasis can lead to serious health problems, such as respiratory failure, recurrent airway collapsing and pneumonia. Bronchiectasis is really important to look out for in young patients who might have cystic fibrosis as they are more prone to this condition.

Most women with bronchiectasis undergo successful pregnancy, however bronchiectasis can predispose to respiratory infection during pregnancy and the mother should be treated with antibiotics.

Management of Bronchiectasis

Bronchiectasis can't be cured; however, with proper care, quality of life is not affected. The underlying principle of therapy is to prevent further scarring and recurrent infection. Being up-to-date with influenza and pneumococcal vaccination is just as important. Bronchodilators and antibiotics may be helpful along with chest physiotherapy. The sooner your doctor can start treating your bronchiectasis and any underlying associated medical conditions, the better the chances are of preventing further damage to your lungs.

Pneumonia

Pneumonia is a common lung infection caused by bacteria, virus, or fungi. Pneumonia and its symptoms can vary from mild to severe. Most patients present with fever, chills, rigors, and excessive tiredness. A chest X-ray is required to make a diagnosis.

The most common cause of bacterial pneumonia in adults is *Streptococcus pneumoniae* (pneumococcus), but there is a vaccine available for this form of pneumonia. There are other bacteria that can cause infection of the lungs that may have atypical features in presentation and are called atypical pneumonias, (also called walking pneumonia); this is caused by bacteria such as *Legionella pneumophila, Mycoplasma pneumoniae, and Chlamydia pneumoniae.*

Patients require an antibiotic for treatment of pneumonia. Treatment depends on the cause of your pneumonia, how severe your symptoms are, and your age and overall health. While most cases of pneumonia may be treated on an outpatient basis, severe cases require hospital admission. Most healthy people recover from pneumonia in one to three weeks, but pneumonia can be life threatening. Pregnant patient may take a little longer to recover from a bout of pneumonia and may require prolonged observation.

The flu virus is the most common cause of viral pneumonia in adults. Other viruses that cause pneumonia include respiratory syncytial virus, rhinovirus, severe acute respiratory syndrome (SARS) and more.

Cold or Flu

Influenza or flu, is a viral infection that attacks the respiratory system: the nose, throat, bronchial tubes and lungs. Flu poses greatest risk to elderly adults, infants and those who have diabetes, chronic heart disease, or lung disease or an impaired immune system. Although the intestinal ailments, such as gastroenteritis (a condition that causes diarrhea), nausea and vomiting are often referred to as flu, they are not. It can be difficult to tell whether you have flu because many of the symptoms are similar to those of a common cold. A cold is a viral infection of the upper respiratory tract (mucus membrane) or the nose, throat and airways to the lungs caused by the Rhinovirus. Flu (fever, chills, rigors, headache, fatigue) usually occur suddenly after an incubation of about one to four days. Flu symptoms last about a week. There are five tests available that your physician can use to diagnose flu as follows:

- Rapid screen for influenza A & B.
- Rapid screen for RSV (respiratory syncytial virus).
- Direct immunofluorescence test of sputum.

- Respiratory culture.
- Respiratory virus panel by polymerase chain reaction (PCR).

Rapid screens for influenza A & B viruses are used during the flu season. These tests give results in short period of time, but they are not as sensitive or able to test for many different viruses as the other tests. In a suspected case of flu, a nasopharyngeal sample is sent for rapid screen. If that is negative, then the sample is usually further tested by either viral culture or respiratory viral panel PCR. A culture can take 2-3 days to get an answer, but a culture is more sensitive and can detect more types of viruses than the rapid tests. The respiratory viral panel by PCR is a molecular test that uses PCR to amplify DNA or RNA from respiratory viruses present in the patient's nasopharyngeal swab or sputum sample. It is the most sensitive test and can detect up to 15 different viruses that can cause infection. Getting a yearly influenza vaccine (unless contraindicated) can greatly diminish your chances of getting the flu.

Differences Between Colds And the Flu

With a cold	With the flu
Symptoms are usually less severe than flu symptoms	Symptoms are usually more severe than cold symptoms
Symptoms develop gradually over a few days.	Symptoms come on quickly and severely.
You rarely have a fever	You almost always have a fever.
You feel sick mostly in your head and nose.	Your entire body feels sick.
Body aches, headaches and pain are usually mild if you have them.	Body aches, headaches and pain are common and can be severe
You may or may not feel tired and weak.	Tiredness and weakness are common.
There is no vaccine to protect you.	You can get vaccine (a shot or the nasal spray) to protect yourself.

Colds generally do not result in serious health problems, such as pneumonia, bacterial infections, or hospitalizations.	The flu can result in serious health problems, such as pneumonia, bacterial infections, or hospitalizations.

Similarities Between The Cold And The Flu

With The Cold	**With The Flu**
Caused by a virus.	Caused by a virus.
Affects the body's breathing system (nose, throat, windpipe and lungs).	Affects the body's breathing system (nose, throat, windpipe and lungs).
Usually goes away on its own.	Usually goes away on its own.
You should contact your doctor if symptoms change or get worse.	You should contact your doctor if symptoms change or get worse. There are antiviral medicines, by prescription, to treat the flu.

*Massachusetts Department of Public Health Updated 2016

Pulmonary Embolism and Deep Vein Thrombosis

A pulmonary embolism (PE) is a blood clot that blocks the blood vessels supplying the lungs. The clot (embolus) most often comes from the leg veins and travels through the heart to the lungs. When the blood clot lodges in the blood vessels of the lung, it may limit the heart's ability to deliver blood to the lungs, causing shortness of breath and chest pain and, in serious cases, death.

Pregnancy leads to a hypercoagulable state with an increased risk for deep vein thrombosis (DVT) and Pulmonary Embolism. Symptoms of a DVT include: calf pain and swelling. The symptoms of pulmonary embolism include: sudden onset of shortness of breath, faster breathing, faster heart rate, and chest pain or swelling of the leg. The presence of a DVT increases the likelihood that the patient has a PE.

Patients that have been diagnosed with pulmonary embolism receive at least 6 months of anticoagulation treatment. Many patients are advised to continue anticoagulation treatment for a longer duration. Treating other risk factors for pulmonary embolism is a critical step in preventing future blood clots. Lifestyle changes such as regular exercise, a heart-healthy diet and smoking cessation are important steps in reducing risk.

Treatment of DVT and PE involves the use of anticoagulants. Since warfarin is contraindicated in pregnancy, low molecular weight heparins (LMWH) are the drug of choice. Patients who have a history of a previous DVT or PE can be managed prophylactically by either empiric treatment with subcutaneous heparin.

Pulmonary Arterial Hypertension

Pulmonary arterial hypertension (PAH) is a progressive disease, which can be life-threatening. The disease is defined by sustained elevations in pulmonary artery blood pressure. This leads to restricted flow in the vessels in the lungs, which makes the right side of the heart work harder. Eventually, the right side of the heart becomes overworked and enlarged.

Advanced pulmonary hypertension patients will have symptoms of shortness of breath, which is first noticed during exercise. As the disease progresses, shortness of breath may occur even at rest. Other symptoms include fatigue, cough, dizziness, chest pain, ankle swelling, and fainting.

Pulmonary hypertension is a contraindication to pregnancy. Normally women with pulmonary hypertension are advised not to get pregnant or are aware of their condition prior to pregnancy. For those patients who choose to get pregnant despite the maternal and fetal risk, prostacyclin analogues are often the treatment. Delivery is done by Cesarean section to avoid blood loss; the greatest risk to the mother is the first month after delivery.

Pulmonary Arterial Hypertension and Pregnancy

PAH affects mostly women, the majority of whom are of childbearing age. Due to the high maternal fetal morbidity and mortality of pregnancy in PAH, pregnancy is absolutely contraindicated and it is recommended to have an early termination. However, with advances in diagnostic modalities and a greater understanding of this disease, many women are electing to keep their pregnancies. If you do have a history of PAH or are diagnosed during pregnancy, please let your physician know immediately as you would be considered a high risk pregnancy, but it may be managed with a multi-physician approach.

Chronic Obstructive Pulmonary Disease (COPD)

Chronic obstructive pulmonary disease (COPD) refers to a condition of the lungs that makes breathing more difficult. It most commonly occurs in people who have a long-standing history of smoking.

COPD is a progressive condition that is characteristically not fully reversible. However, if you get appropriate treatment and make lifestyle changes, you can slow the damage and improve your lung function. This will lead to you coughing less, breathing more efficiently and feeling better.

Chronic = always present to some degree
Obstructive = blockage in the airways
Pulmonary = lungs
Disease = illness

The most common syndromes are chronic bronchitis and emphysema. Please note that these are often found in combination, so a patient diagnosed with COPD can have both emphysema and bronchitis, simultaneously.

Chronic Bronchitis

Chronic bronchitis causes inflammation in the airways and overproduction of mucus. Inflammation narrowing and mucous plugging can clog the airways, making it difficult to breathe. Excess mucus is a common feature of this disease. Chronic bronchitis occurs as a result of constant irritation of the airways from cigarette smoke or other inhaled environmental pollutants. A constant cough develops in an attempt to rid the lungs of the excess mucus. Although we all cough, chronic bronchitis patients have a cough with excessive mucous, which occurs for at least three months out of a year for two consecutive years.

Emphysema

Emphysema causes inflammation and damage to the fragile walls of the air sacs in the deepest part of the lungs. This structural damage causes the air sacs and small airways of your lungs to collapse as you breathe out. Constant irritation from smoke or other pollutants damage the airways causing them to become narrow and thereby limiting the passage of air leaving the lungs. As a result, air becomes trapped inside the alveoli causing them to distend and eventually rupture. These damaged alveoli provide less surface area for the exchange of oxygen and carbon dioxide, resulting in a less efficient way of breathing.

Worldwide, cigarette smoking is the most overwhelmingly common cause for emphysema. Other exposures which can cause emphysema include occupational dusts and indoor air pollution from biomass fuel often used in cooking. Whereas chronic bronchitis is a disease of the airways, emphysema is a disease of the air sacs, or alveoli.

Alpha-1 antitrypsin deficiency

Alpha-1 antitrypsin deficiency is a hereditary disorder characterized by reduced levels of alpha-1 antitrypsin and results in emphysema in an earlier age group.

Alpha-1 antitrypsin is a blood protein that is produced in the liver and its main function is to protect the lungs so they can work normally. After the liver releases it into the bloodstream, alpha-1 goes into the body tissues and protects the tissues from being digested by enzymes released from inflammatory cells, such as white blood cells. In normal lungs, alpha-1 antitrypsin protects lung tissue by trapping and destroying these enzymes from inflammatory cells before it has a chance to cause damage. Certain gene mutations can cause an abnormal form of alpha-1 antitrypsin that gets hung up in the liver and cannot enter the blood stream. This results in low levels of alpha-1 antitrypsin in the blood.

When the lungs do not have enough alpha-1 antitrypsin, lung tissue is particularly vulnerable to digestion enzymes secreted by inflammatory cells, since the lungs are continuously exposed to toxins in the environment. This leads to lung disease, most commonly emphysema.

Patients with emphysema are treated with inhaled steroids and bronchodilator therapies.

Tuberculosis

Tuberculosis (TB) is a communicable infection (can be transmitted from person to person) that usually affects the lungs. It is spread by airborne droplets when an infected person coughs or sneezes. It is caused by a bacterium called *Mycobacterium tuberculosis*. There are two types of tuberculosis infections. One is dormant, which is known as latent tubercular infection (LTBI) and the other is active tuberculosis infection. At the time of diagnosis, people with active TB infection usually have a variety of symptoms such as low-grade fever, constant cough with sputum (phlegm), night sweats, and unintentional weight loss.

Classifying TB

- Active TB describes an ongoing infection in which a person develops symptoms and has a positive (abnormal) result on a test for TB.
- Latent tubercular infection (LTBI) occurs when a person with no symptoms has a positive result on a TB skin or blood test. This suggests that the person was infected with TB in the past but the bacteria are in a dormant or inactive state. Persons with latent TB cannot spread the TB bacteria to others.

Risk Factors for TB

- Traveling to or living in countries where tuberculosis is endemic
- Working in a health setting
- Living in overcrowded and poorly ventilated residences
- A deficient or weakened immune system, such as in people with diabetes or HIV/AIDS

Evaluation for suspected TB

- Tuberculin skin test (also called PPD or purified protein derivative). In response to this injection, if a person has been infected with TB, immune cells will indurate (harden) the area surrounding the injection site. The area of induration is measured 48 to 72 hours after injection and used to determine the likelihood of TB infection.
- Chest x-ray may be done to distinguish between active and latent TB.
- A blood test may be done to check for cytokines (substances released by immune cells) that are unique to TB infections. QuantiFERON-TB Gold is an in vitro (blood sample) diagnostic test that aids in the detection of tuberculosis infection. QFT is an interferon γ (IFNγ) release assay (IGRA) which measures the cell-mediated response to specific TB antigens in whole blood which aid in determining if you have been exposed to the tubercular bacilli.

Treatment

- Anti-TB antimicrobials (medicines that kill microorganisms or interfere with their growth) are used to treat acute tuberculosis.
- Treatment usually lasts for 6 months and requires close monitoring by an infectious diseases specialist or other specialist.
- Patients diagnosed as having LTBI may be given medications to kill dormant bacteria and prevent the development of active TB.
- In latent tuberculosis, infections treatment is generally held until delivery of the baby.

Tuberculosis and Pregnancy

If you are pregnant and you are coughing up mucous for more than three weeks, your doctor will ask you to do a sputum test. This may be followed by a chest X-ray, skin test, or a blood test. It's important to have any tests that are recommended. The impact that undetected and untreated TB will have on your baby is greater than any potential harm that an X-ray might have on your baby.

If you are diagnosed to have TB during pregnancy, you may be started on treatments. Most of the medications are safe to be taken during pregnancy, but in most cases treatment is withheld until you deliver.

If you have TB during pregnancy and if you don't diagnose and treat it early enough, there is an increased risk of:

- Having a miscarriage
- Your baby being born prematurely
- Your baby being born with low birth weight
- Rarely, your baby may be born with TB.

Your baby can catch TB from you at birth only if you have active TB in the lungs and have not begun treatment.

Pulmonary Edema

Pulmonary edema in pregnant women is life threatening to both the mother and the child. Pulmonary edema occurs when fluid enters the lung faster than it can be removed leading to accumulation in the lung. Pregnant women can develop pulmonary edema during delivery if they have a congenital or rheumatic heart disease and systemic hypertension (high blood pressure). Sometimes drugs used to reduce the rate of uterine contractions, known as tocolytics, when administered during delivery, can cause pulmonary edema. Pregnant women that develop a condition called pre-eclampsia and eclampsia are prone to develop fluid overload and further leading to pulmonary edema. Most patients who develop pulmonary edema will have to be admitted to the hospital and treated.

When 'I' is replaced By 'we'

Even
'illness'

Becomes
'Wellness'

4 | Diagnosing your Disease

"If you get a diagnosis, get on a therapy keep a good attitude and keep your sense of humor."

Teri Garr (b. 1944)

In the following section, we will discuss information on different tests that your physician may order to help diagnose your heart and lung disease.

Spirometry

Spirometry is a routinely performed breathing test that assesses how well your lungs function. During this test, you will be asked to blow into a tube which is connected to a machine. This machine, or a spirometer, measures how much air your lungs can hold (forced vital capacity or FVC) and how fast you can blow air out (forced expiratory volume in one second or FEV_1). Along with a doctor's examination, spirometry can help determine the severity of your COPD.

Pulmonary Function Testing (PFTs)

Pulmonary function testing is a commonly used test, ordered by the physician to evaluate the function of lungs. This testing involves the use of machinery to perform breathing tests to measure the size of lungs, the state of the airways of the lungs, and the ability of the lungs to exchange oxygen and carbon dioxide.

Patients are asked to withhold long-acting lung medications for 12 hours and short-acting lung medications for 4 hours prior to testing. Patients are asked not to engage in vigorous physical exercise at least 4 hours prior to testing. Patients may be asked for a sample of blood to be drawn for measuring the amount of the oxygen and carbon dioxide of the blood. This is done by a certified Respiratory Therapist using local anesthetic called lidocaine to numb the area over the artery in which the needle is inserted. Patients may be asked to inhale medication to open the airways in the lungs. This allows the doctors to determine how well the lungs respond to this type of medication.

Ankle Brachial Index

A test called an ankle brachial index or ABI is used to diagnose peripheral artery disease (PAD). Health care providers compare the blood pressure in the ankle with that in the arm. Lower blood pressure in the lower part of the leg compared with the pressure in the arm may indicate PAD.

Cardiac catheterization

Left heart cardiac catheterization is used in conjunction with other tests. A small tube is inserted into an artery and guided into the blood vessel of the heart. It helps to locate blockages in the heart vessels.

Right Heart Catheterization

Here a small tube is passed in the right side of the heart to measure the pressures in the pulmonary artery. This is the gold standard test for diagnosing pulmonary artery hypertension.

Chest X Ray

A chest X-ray shows the size and shape of the heart and can also show congestion in the lungs and diseases of the pleura.

CT (Computed Tomography) Scan

A CAT scan uses special scanning techniques to provide images of the lungs. You may be given a contrast prior to the test; this study is then called as a CT Angiography. This test allows your physician to evaluate your pulmonary arteries to look for any clots in them.

Echocardiogram

An echocardiogram uses high-frequency sound waves (ultrasound) to produce images of the heart and blood vessels. Results indicate whether the heart is pumping blood correctly. It also allows us to estimate the pressure in the pulmonary arteries. This is good screening test when your health care provider wants to rule out pulmonary artery hypertension. A stress echocardiogram uses either exercise or medication and ultrasound to provide images of the heart and blood vessels under stress.

Electrocardiogram (EKG, ECG)

An electrocardiogram, also called an ECG or EKG, provides information on heart rate and rhythm and shows whether there has been any damage or injury to the heart muscle.

Holter Monitoring

A Holter monitor is a small, portable machine that records the heart's electrical activity. The person wearing the monitor keeps track of symptoms and activities for the evaluation period. Readings on the machine are compared with the symptoms.

Magnetic Resonance Imaging (MRI)

MRI uses special scanning techniques to provide images of body tissues. MRA (magnetic resonance angiography) uses MRI to examine blood vessels.

Nuclear Medicine Lung (VQ) Scans Information for Patients during Pregnancy

A nuclear medicine Ventilation and Perfusion lung (VQ) scan test is performed to see if a person has a blood clot in their lungs (called a pulmonary embolus).

The test is usually done in two parts. During the first part, the patient wears a mask over their nose and mouth so that they can breathe in a small amount of a radioactive gas which is used to obtain images of the lungs (sometimes this part of the test can be omitted). For the second part of the exam, a small amount of a radioactive material is injected into a vein in the patient's arm. After the injection, images of their lungs are obtained. The test does expose the patient to some radiation and also to the fetus during pregnancy.

The effects of high doses of radiation on an unborn baby (fetus) have been associated with birth defects, growth retardation, and abnormal brain development. However, the minimal dose the feus receives during this procedure has not been associated with fetal anomalies (birth defects) or fetal loss. Complications have only been seen with radiation doses much higher than those used for this test.

This study helps doctors to find out if the patient has a blood clot in their lungs which is a potentially life-threatening condition. Patients are advised to discuss with their physician about the benefits and risks of this test before proceeding.

Radiation Exposure due to CT Scan and X-Rays

Routine X-rays are usually postponed until after delivery, just to be on the extra-safe side. The risks of X-rays during pregnancy are very low and can be easily made even lower. What's more, just a case in point, a typical diagnostic X-ray of any kind rarely delivers more radiation that you'd get from spending a few days in the sun at the beach.

- Always inform the doctor ordering the X-ray and the technician performing it that you're pregnant, even if you're pretty sure they know
- A lead apron should be used to shield our uterus and a thyroid collar should protect your neck.

5 | Pulmonary Medications and Pregnancy

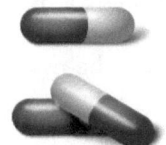

"The best doctor gives the least medicine."

Benjamin Franklin (1706-1790)

There are many different types of prescribed medications that may help you breathe easier. Your doctor may have already prescribed some of these for you. This section is designed to help you understand why you're taking a particular medication, how the various medications work, and if there are any possible side effects. You should always tell your doctor if you are taking any other medications (which should only be used under the direction of your physician). This is particularly true of narcotics, sleeping pills, or tranquilizers. Before taking any medication, please make sure that you discuss with your pharmacist and physician regarding the safety of the drug. Below is a classification system of medicine set forth by the Food and Drug Administration (FDA).

FDA Pharmaceutical Pregnancy Categories

Category A	Adequate and well controlled human studies demonstrate no risk
Category B	Animal studies demonstrate no risk, but no human studies have been performed OR Animal studies demonstrate a risk, but human studies have demonstrated no risk

Category C	Animal studies demonstrate a risk, but no human studies have been performed. Potential benefits may outweigh the risks
Category D	Human studies demonstrate a risk. Potential benefits may outweigh the risk
Category X	Animal or human studies demonstrate a risk. The risks outweigh the potential benefits

Preventing Medication Errors

Know your medication: Keep a list of all your medications, how much you take and when you take. Include over the counter medicines, vitamins, supplements, and herbs. Take this list to all your doctor visits. Keep this list current at all the times. Never take medications prescribed for others. Ask your healthcare provider if you do not know the answer to these questions. Please ask your doctor or pharmacist the following:

- Why am I taking this medication?
- What are the common problems to watch for?
- What should I do if they occur?
- When should I stop this medicine?
- Can I take this medication with the other medicines on my list?

Expectorants and Mucolytics (Decongestants)

These drugs are designed to increase removal of respiratory tract fluids and help liquefy mucus that is sticky or thick. They're available in liquids, elixirs, drops, liquid capsules, and liquid for inhalation only. Some examples are water, Guaifenesin, and Iodide Humibid.

Drinking fluids is the best way to liquefy mucus. Expectorants should not be combined with antihistamines such as Benadryl® or cough suppressants like codeine or dextromethorphan.

Aspirin

Individuals with asthma or asthma symptoms might find that aspirin causes excessive shortness of breath and wheezing. If this occurs, discuss a substitute with your physician. There are a variety of aspirin substitutes on the market today.

Antibiotics

Antibiotics help fight existing bacterial infections and help prevent future severe infections. There are hundreds of different types and your physician will make the decision on which one is the best for your specific situation. Antibiotics are available in tablets, capsules, liquids, and injections in the vein. Possible side effects of antibiotics may be upset stomach or allergic reactions; others include nausea, vomiting or diarrhea. There may be other side effects as specified by your doctor.

Bronchodilators

Bronchodilators relax muscles around the airways to keep them open, making it easier to breathe. They're available in tablets, elixirs, inhalants, suppositories, and injections. Examples of these medications include: Albuterol (Proventil®/Ventolin®), Serevent®, Metaproterenol (Alupent®; Metaprel®) Terbutaline (Brethine; Bricanyl), Theophylline, Brovana® (Arformoterol) and Xopenex® (Levosalbutamol). Some of these medications may be available in inhaler form; however, some may be available in nebulizer form as well. Possible side effects are increased heart rate, anxiety, nervousness, shakiness, headaches, upset stomach, heartburn, loss of appetite, sleeplessness, and sweating.

Steroids

Steroids decrease airway swelling and inflammation. They help decrease wheezing by allowing the airways to remain relaxed and open. They're available in tablets, inhalers, and intravenous form. The oral steroids are prednisone and methylpednisolone (Medrol®). The inhaled steroids are

QVar® (beclomethasone), Aerobid® (Flunisolide), Azmacort® (Triamcinolone), Flovent® (Fluticasone), and Pulmicort® (Budesonide)

Possible side effects of oral steroids are weight gain and puffiness due to fluid retention; stomach ache; fragile bones; easy bruising of skin; lowered resistance to infections. Inhalers may cause dry mouth, hoarseness and oral-dental fungal infections. It is important to remember not to discontinue or change the dosage of steroids on your own. Changes should only be made by your physician and usually will involve a gradual taper. Take oral steroid medications with food, milk, or antacids to ease possible stomach problems. After using inhalers, rinse your mouth to avoid any irritation and infection. These medication are generally considered safe but please ask your physician before use.

Combination of Steroids and Bronchodilators

There are options that combine Steroids and Bronchodilators into one inhaler or nebulizer. This allows for an anti-inflammatory effect as well as a treatment to open the airways in one medication. Some examples are Advair® (fluticasone/salmeterol), Symbicort® (budesonide/formoterol), and DuoNeb® (ipratropium bromide/salbutamol)

Diuretics (water pills)

Since diuretics help rid the body of excess fluids, they may cause mucus membranes to dry out. Thick, sticky mucus is harder to cough up and creates an environment for bacterial growth. If you are taking water pills for another condition, you should contact your physician. It is possible your doctor will supplement your potassium while you are on water pills.

Anticoagulants (blood thinners) for Venous Thromboembolism

If someone is ever diagnosed with a clot in their lungs called a pulmonary embolism or DVT, they will be initiated on anticoagulant therapy. Most commonly in pregnancy, the medications that may be used safely are subcutaneous heparin and low molecular weight heparins (Lovenox ®).

Immunizations In Pregnancy

Since infections of various sorts can cause pregnancy problems, it's a good idea to take care of all necessary immunizations before conceiving. Most immunizations using live viruses are not recommended during pregnancy, including the MMR (measles, mumps and rubella) and varicella (chicken pox) vaccines. Other vaccines, according to the CDC, shouldn't be given routinely but can be given if they're needed. These include the hepatitis A pneumococcal vaccine. You also can be immunized safely against hepatitis B when you're expecting. In the must-have department: The CDC recommends that every woman who is pregnant during flu season (generally October through April) receive a flu vaccine and that every pregnant woman get the Tdap vaccine (which offers protection from diphtheria, tetanus, and pertussis) between 27 and 36 weeks of pregnancy, regardless of when she was last vaccinated with the Tdap or Td vaccine.

Tetanus Toxoid Vaccine

Tetanus infection generally occurs through contamination through an open wound that involves a cut or deep puncture. If tetanus develops, a person may develop muscle spasms in the jaw and elsewhere in the body. Tetanus may be prevented by a tetanus vaccine. Most individuals have already been vaccinated at one point in their lives but it is recommended that you receive a tetanus booster every 10 years.

For more information about which vaccines are safe during pregnancy, check with your practitioner (See Appendix 2).

How to Use Your Inhalers

How to use a Metered Dose Inhaler (MDI) (Albuterol, Proventil®)

Step 1: Remove the cap from the inhaler.

Step 2: Shake the inhaler well for 5 seconds.

Step 3: Hold the inhaler firmly by placing your index finger on top of the canister, and thumb on the bottom of the mouth piece.

Step 4: Sit strait or stand and slightly tilt your head back.

Step 5: Exhale away from the inhaler.

Step 6: Put the inhaler in your mouth. Press the inhaler and start breathing in at the same time. Take a slow and deep breath.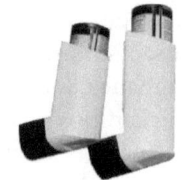

Step 7: Hold your breath for 10 seconds. Exhale slowly though your mouth or nose.

Step 8: Repeat steps 2 to 7 after 30 seconds, if another dose is required. If you are using a corticosteroid medication, rinse your mouth after all doses are complete.

How to use a Diskus Inhaler (Advair®, Flovent®)

Step 1: Check the dose counter to see the number of doses remaining.

Step 2: Hold the inhaler properly in both the hands. Open the inhaler by using the thumb grip slide. You'll hear a click.

Step 3: Hold the inhaler horizontally. For loading the dose, slide the lever downwards. You will hear a click.

Step 4: Exhale away from the inhaler.

Step 5: Place the mouthpiece in your mouth. Take a quick and deep breath. Hold your breath for 10 seconds. Exhale slowly through your mouth or nose.

Step 6: If you are using "Flovent Diskus" and have been advised to take another dose, then repeat the steps from 3 to 6, after 30 seconds. If you are using a corticosteroid medication, rinse your mouth after all doses are complete.

How to use a Handihaler (Spiriva®)

Step 1: Open the dust cap by pulling upward and then open the mouth piece.

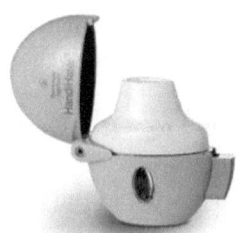

Step 2: Keep one end of the capsule in the center chamber.

Step 3: Close the mouthpiece firmly until you hear a click and leave the outer cap open.

Step 4: Press piercing button once. You'll hear a click. Doing so makes holes in the capsule and easily releases the medicine, when you inhale.

Step 5: Exhale away from the inhaler then put the mouthpiece in your mouth and take a quick and deep breath. You'll hear the capsule rattle.

Step 6: Hold the Handihaler horizontally. Keep the mouthpiece in your mouth and take a quick and deep breath. You'll hear the capsule rattle.

Step 7: Hold your breath for 10 seconds. Exhale slowly through your mouth or nose.

Step 8: Breathe in through the mouthpiece once again. This ensures that you get your complete dose of medicine.

Step 9: Open the mouthpiece, remove the used capsule and dispose it. Do not keep capsules in the Handihaler. For storing the Handihaler. Close the mouthpiece and dust cap.

How to use a Twisthaler (Asmanex®)

Step 1: Hold the Twisthaler vertically and rotate the cap towards the left to open it.

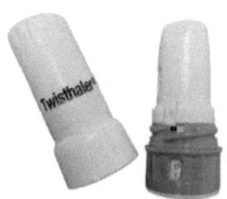

Step 2: When you lift the cap it will load the medication dose and move the counter.

Step 3: Exhale away from the inhaler, then keep Twisthaler in your mouth.

Step 4: Take a quick deep breath and hold for 10 seconds. Then exhale.

Step 5: Immediately put the cap back on the inhaler and rotate to close until you hear a click.

How to use a Respimat (Combivent, Spiriva)

Step 1: Hold the Respimat vertically.

Step 2: Turn the clear base in the direction of the arrow until you hear a click.

Step 3: Open the cap and sit or stand straight up.

Step 4: Slightly tilt your head back and exhale away from the inhaler.

Step 5: Place Respimat in your mouth and take a slow deep breath while simultaneously pressing the dose dispensing button.

Step 6: Hold your breath for 10 seconds and exhale slowly through your mouth or nose.

6 | Exercise During Pregnancy

"Walking is man's best medicine."

Hippocrates

Exercise during pregnancy is often considered taboo or wrong, however the exact opposite is true. An exercise regimen is more than just physical therapy; it may also benefit you mentally and emotionally.

Exercise Leads To

- Improved activities of daily living
- Nutritional awareness
- Educational awareness
- Shortness of breath reduction

Emotional and Physical Rewards of Exercising

- Improves blood flow circulation and improves oxygen delivery to the body
- Decreases stress hormones
- These changes help to ward off depression

Exercise Training

Before beginning an exercise regimen during pregnancy, please consult with your physician. As your pregnancy progresses, you may find some exercises more difficult than others. However, this chapter will outline general considerations of exercise during pregnancy.

Healthy pregnant women need at least 2½ hours of exercise each week. This is about 30 minutes each day. If this sounds like a lot, don't worry. You don't have to do it all at once. Instead, split up your exercise by doing something active for 10 minutes three times each day. Physical activity is good for you as it can:

- Keep your heart, body and mind healthy
- Help you feel good and find the extra energy you need
- Help you stay fit and gain the right amount of weight during pregnancy
- Ease some of the discomforts you might have during pregnancy, like constipation, backaches, trouble sleeping and varicose veins (swollen veins)
- Prevent health problems like preeclampsia and gestational diabetes
- Help your body get ready to give birth
- Help reduce stress

What types of activities are best during pregnancy?

If your provider says it's OK for you to exercise, pick activities you think you'll enjoy. Some hospitals and health clubs offer aerobics and yoga classes just for pregnant women. Or try things you can do with your partner or friends, like walking or dancing.

Swimming is a great activity for pregnant women. The water supports the weight of your growing body, and moving against it keeps your heart rate up.

If you exercised before you were pregnant, it's usually safe to continue your activities during pregnancy. Check with your provider to make sure. As your pregnancy continues and your belly gets bigger, you may need to change some activities or ease up on your workout. If you didn't exercise before you were pregnant, start slowly. Try to build up your fitness little by little.

Breathing Exercises in Pregnancy

Pursed Lip Breathing

The goals of breathing exercises are to:

- Improve the function of the diaphragm.
- Control respiratory rate and decrease the work of breathing.
- Assist in relaxation and thereby alleviate dyspnea.
- Increase the strength, coordination and efficiency of breathing patterns.
- Prevent or reverse atelectasis (lung collapse).
- Mobilize and maintain mobility of the chest wall.

Pursed-Lip Breathing

Pursed-lip breathing is effective in reducing the respiratory rate and relieving dyspnea. It has been suggested that this method of breathing may improve ventilation and oxygenation.

- Relax your neck and shoulder muscles.
- Inhale slowly through your nose for at least 2 counts.
- Pucker your lips as if to blowout a candle.
- Exhale slowly and gently through your pursed lips for at least twice as long as you inhaled.

You can practice breathing this way anytime, anywhere. If you're watching TV, practice during the commercials. Try to practice several times a day. Over time, pursed-lip breathing will feel natural.

- Use pursed-lip breathing to prevent shortness of breath when you do things such as exercising, climbing stairs and bending or lifting.
- Breathe out during the difficult part of any activity, such as when you bend, lift, or reach.
- Always breathe out for longer than you breathe in. This allows your lungs to empty as much as possible.
- Never hold your breath when doing pursed-lip breathing.

Physical Exercises

Stretching and Strengthening Exercises

Stretching Flexibility Program - Unless otherwise noted, each stretch should be performed 3 times a week and held for 30 seconds. Refrain from bouncing or jerky movements. A rest period of 5 seconds is sufficient between each repetition. While performing these stretches, you may experience some minor discomfort, or even a mild "burn", within the muscle. No joint pain should be experienced. If any pain is felt in the joint, discontinue the exercise and consult your therapist.

Flexibility exercises - These exercises should be performed 5 times per week prior to your lower extremity exercises (treadmill, stationary-bike, or free walking).

Lower Body Exercise Endurance Program - Stretching exercises should be performed prior to your lower body work out and a cool-down immediately following exercise for approximately 3 minutes. Begin exercising for 10 minutes daily, then increase to 20 minutes. After you perform 30 minutes of continuous exercise, increase the intensity. Your goal is to complete 30 minutes of continuous exercise. If you are performing free walking, remember to walk with your arms hanging loosely, your chest and shoulders relaxed. Lower body exercises should be performed 5 times a week.

Upper Body and Strength Training Program - Arm raises seated in a chair should be performed initially for 10 minutes without weights. When you are able to complete 10 minutes of continuous exercise, add ½-1.0 lb weights to your routine. Therabands (elastic bands) or free weights can be used to strengthen your body. You should perform 5-6 different exercises for both the upper and lower body. Each exercise should be performed initially only 10 times per set. Gradually increase to 20-30 times per set and then increase intensity. All strengthening exercises should be done in a slow, controlled manner. Avoid any explosive or sharp movements, as this can strain your muscles and your breathing. Upper body and strength training should be performed 3 times a week.

Below is a sample physical exercise regimen which can help. Keep in mind that your exercise plan should be reviewed by your physician to ensure that you are not putting yourself at any risk of injury. Durations and weight limits should be decided after a consultation with a physical therapist, respiratory therapist and your physician.

Do NOT hold your breath during these exercises. Inhale through your nose and exhale through pursed lips as you exert force.

Stretching Exercises

The following exercises are designed to gradually increase the strength of various muscle groups. Strengthening of your muscle will help with your overall health, but, most importantly, it will help streamline your lung function. All exercises should be initially performed under supervision. Weightlifting exercises should be performed using a comfortable amount of weight; remember not to exert yourself. They may be repeated in sets of 3 or whatever you feel comfortable doing.

Cervical Spine and Neck

Flexibility of Neck

Exercise 1

- Place hand on same side shoulder blade
- With other hand gently stretch head down and away
- Hold for 3 seconds

Exercise 2

- Gently grasp side of head while reaching behind back with other hand
- Tilt head away until a gentle stretch is felt
- Hold for 3 seconds

Flexibility: Neck Retraction

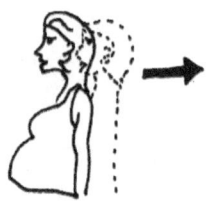

- Pull head straight back keeping jaw and eyes level
- Hold for 3 seconds

Cervical Spine - Phase I: Shoulder Shrugs

- Shrug shoulders up and down, forward and backward
- Hold for one second

Flexibility: Corner Stretch

- Standing in corner with hands at shoulder level and feet a comfortable distance from the corner, lean forward until a comfortable stretch is felt across chest
- Hold for 2 seconds

Phase II: Resistive Shoulder Shrugs

- With surgical tubing/dumbbells, shrug shoulders up and down, forward and backward

Back - Lumbar Rotation in Sitting

- Gently rotate trunk side to side in a small, pain-free motion

Hip and Knee - Standing Hamstring Stretch

- Pull knee toward chest until easy stretch is felt

Calf Stretch

- Straighten leg out in front of you
- Bring your toes back toward your knee as you push your heel forward

Hip and Knee - Gastrocnemius Stretch

- Keeping back leg straight, with heel on floor and turned slightly outward, lean into wall until a stretch is felt in calf

Hip and Knee - Soleus Stretch

- Keeping back leg slightly bent, with heel on floor and turned slightly outward, lean into wall until a stretch is felt in calf

Quadriceps Stretch

- Pull heel toward buttock until a stretch is felt in front of thigh

Shoulder - Range of Motion Exercises (Self-stretching activities)

External Rotation (alternate)

- Keeping palm of hand against door frame and elbow bent at 90 degrees, turn body from fixed hand until a stretch is felt

Cervical Spine - Lower Cervical/Upper Thoracic Stretch

- Claps hands together in front with arms extended. Gently pull shoulder blades apart and bend head forward

Chest/Biceps Stretch

- Lace fingers behind back and squeeze shoulder blades together. Slowly raise and straighten arms

Strengthening Exercises - Biceps

- Stand erect or sit in a chair
- Hold dumbbells at arm's length, palms in
- Keep back straight, head up, hips and legs locked
- Curl dumbbell in right hand with palm in until past
  thigh, then palm up for remainder of curl to shoulder
- Keep palms up while lowering until past thigh, then turn palms in
- Keep upper arms close to side
- Do a repetition with right arm, then curl left arm or do both arms at the same time

Triceps

- Stand erect or sit in a chair
- Hold dumbbell in right hand; raise overhead to arm's length, upper arm close to head
- Lower dumbbell in semicircular motion behind head until forearm touches biceps
- Return to starting position and repeat with left arm

Shoulder Flexion

- Start with arm at side.
- Lift your arm toward the ceiling, keeping the arm straight

Shoulder - Progressive Resistive Exercises - Abduction (Standing)

- Raise arms out from body

Knee Lifts

- Repeat desired number of times

Seated Knee Extension

- From a seated position, slowly straighten leg. Slowly return to start position

Knee Flexion

- Standing, bend knee up as far as possible
- Hold for 2 seconds

Hip Flexion

- Hip flexion with support for balance
- Hold for 2 seconds

Heel Raises

- Repeat desired number of times

Hip Extension

- Hip extension with support for balance
- Hold for 2 seconds

Hip Abduction

- Bring leg out to side
- Hold for 2 seconds

Theraband Exercises

Shoulder Flexion

- Sit or stand on firm surface with the Theraband held at hip or waist height
- Point thumb toward ceiling. With elbow straight, raise one hand toward ceiling
- Hold. Return to start position

Chest Pull

- Sit or stand with feet shoulder width apart
- Loop the front of body with elbows slightly bent
- Pull the Theraband outwards, across chest
- Hold. Return to start position

Shoulder Extension

- Grasp the Theraband in palms with arm above head
- Point thumb toward ceiling. Lower arm
- Hold. Return to start position

Seated Row

- Assume long sit position with a straight back
- With the Theraband looped under both feet, hold each end of the theraband with elbows straight
- With arms close to the sides of the body. Pull arms/elbows back
- Hold and lower slowly. Return to start position

Neck Exercises

Bend your neck forward (Flexion) and backwards (extension) in full range. Hold for a while, take it back to normal. Repeat 5 times.

Keep your hand together, rotate the shoulder both clockwise and counterclockwise, and take it back to normal.

Bend your neck sideways as shown. Hold it for a while. Repeat 5 times.

Elevate the shoulder for a while. Bring it to normal. Repeat 5 times.

Using hand as a resistance forcing flexion & extension, hold for a while. Repeat 5 times.

Use a thin, soft pillow or dog bone pillow which you can make by removing some of the cotton stuff from the middle portion.

Yoga and Medication

Certain activities provide a good workout, release tension and decrease anxiety while promoting health benefits. Yoga, a 5000 year old Indian practice, is one such activity and is the best-known mind-body exercise. It involves a series of sitting and laying down postures that help along with coordinated breathing and meditation techniques. Another such technique is Tai Chi, a Chinese method of slow body movement that promotes relaxation. Pilates, like yoga, concentrates on breathing while strengthening the body's core muscles. Meditation can help reduce stress and anxiety, while improving quality of life. It is recommended that you get involved in such activities as part of your therapy.

Stress Reduction Techniques - Relaxation and Psychophysical Techniques

The close association between dyspnea and anxiety is well known. If dyspnea is escalated and reinforced by anxiety, as many of you have experienced, a strategy must be developed to reduce the intensity and distress of dyspnea. There are a variety of different techniques that will help you relax during occasions of severe dyspnea and/or anxiety, as well as to decrease stress in daily life. Progressive muscle relaxation is a widely used technique. Some common components of most relaxation techniques include the following:

- A quiet environnent
- A confortable position
- Loose, non-restrictive clothes
- Adoption of a passive attitude

Progressive Muscle Relaxation

- Close your eyes
- Perform 2 large cleansing breaths followed by your pursed-lip breathing
- With your eyes closed, try to visualize your favorite place to visit. Visualize in your mind the environment, the scenery, the smells and the colors. Try to get in touch with how you feel when you are there. Calm, relaxed and peaceful. Your breathing will begin to slow; you will take slower and deeper breaths.
- Perform slow abdominal breathing with a deep inhalation and a slow exhalation through pursed lips.
- Optional: follow the above mentioned steps followed by systematic tensing then relaxing every part of the body including feet, arms, legs, chest, face, eyes, shoulders, etc., concentrating on each muscle as the tension and relaxation is performed.

7 | Health Maintenance

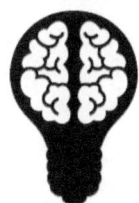

"To keep the body in good health is a duty... otherwise we shall not be able to keep our mind strong and clear."

Buddha (563 BC – 483 BC)

While your doctor is providing you the best way to manage your disease medically, ultimately you are in charge of your own health. In this section, you'll find practical advice on how to handle your health and other suggestions to make life easier, healthier, and more positive.

See your doctor regularly

Even if you're feeling fine, stick to your appointment schedule. Schedule your next appointment while you're at the office, so you won't forget. When you visit your physician, in addition to your prescribed medications, please be sure to inform them of any vitamins, herbal supplements or alternative medicines you may be taking. It is also important to notify your physician of any allergies to food or medications. Common food allergies may be to peanuts, milk, soy, nuts from trees, eggs, and wheat.

Take your medications as prescribed

Bring an updated list of your medication to be reconciled with your chart on every visit. Provide your physician with a mail order pharmacy number as well as a local pharmacy. Before you leave your doctor's office, make sure your refills have been ordered. You may have more than one medical condition that must be considered when making a dietary plan, so always talk with a healthcare provider or registered dietician before making changes in your diet. Set up a system that will help you remember to take your medications at the appropriate times. On the day of an appointment, schedule time in to take your medications.

Track your condition and symptoms

Prior to coming to your doctor visit, you should also write down any questions or topics you want to discuss with your doctor. In addition, please keep a log of your blood pressure, sugars, and oxygen saturation. Make it a point to discuss all of your current and previous lab work. Record dates of all previous diagnostic testing such as echocardiography, stress test, and radiological testing.

Quit Smoking
One of the best ways you can make a positive change in your heart and lung condition is to stop smoking (See Chapter 10). If possible, cut down on the irritants in the air where you work and live. For example, by avoiding hair spray or other aerosol sprays.

Be active and get stronger
Talk to your doctor about what activities are appropriate to do. With regular exercise, you may find an improvement in your symptoms, appetite; sleep patterns and your overall sense of well-being. Regular physical activity reduces your risk of heart disease and stroke. It also helps you reduce or control other risk factors such as high blood pressure, high blood cholesterol, excess body weight and diabetes. But the benefits don't stop there. You may look and feel better if you remain active. You will also become stronger and more flexible, have more energy and reduce stress and tension. The time to start is now.

Exercise regularly
Walking is an excellent activity. Start walking at a slow, comfortable pace for a short period of time (try 5 to 10 minutes) three to five days each week. When you're able to walk the entire time without stopping to rest, you can increase your walking duration by 1 to 2 minutes each week. A sign of a healthy lifestyle is taking 10,000 steps a day. You can monitor your steps per a day by acquiring a pedometer which is available at most sporting or health store. Choose activities you enjoy. Pick a starting date that fits your schedule and gives you enough time to begin your program, like a Saturday.

Some tips to help you exercise are:

- Wear comfortable clothes and shoes.
- Start slowly - don't overdo it.
- Try to exercise at the same time so it becomes a regular part of your lifestyle. For example, you might walk every day (during your lunch hour) from 12:00 to 12:30 or start each morning with stretching and strength training.

- Drink lots of water before, during, and after each exercise session.
- Note the days you exercise and write down the distance or length of time of your workout and how you feel after each session. You may also want to note if your muscles are tired the next day.
- If you miss a day, plan a make-up day. Don't double your exercise time during your next session.

Conserve your energy - Though exercise is a vital part of therapy, it is important to not to overexert yourself as this may actually make your breathing worse.

Have a positive attitude - Try not to compare yourself with others. Your goal should be your own personal health and fitness. Think about whether you like to exercise alone or with other people, outside or inside, what time of day is best, and what kind of exercise you most enjoy doing.

- Join a support group
- Join an exercise class
- Exercise with friends or family to help motivate you

Relax - When you feel stressed, consider a stress reduction strategy as discussed later in the chapter.

Breathe better air - It is important to stay away from cigarette smoking as well as second hand smoke, irritants, mold or any other respiratory triggers. Use an air conditioner and change the air filters frequently to keep the air less humid, cleaner, and more comfortable to breathe.

Vaccinations - It is very important to stay up to date on your vaccinations. For more information on vaccination, please discuss with your physician (See Appendix 2).

The ABC's of Better Health during Pregnancy

- **A**lways be positive! – Each day is a new day and you can make an effort to improve your breathing.
- **B**reathe – Your breathing is something that can be improved with a little work and exercise. Follow the guidance of your physician.
- **C**oncentrate – Focus on the process of your health during pregnancy, so you can gain a better understanding of it.
- **D**iary – Keep a diary of things that make you feel comfortable. Exhale - During any activity in which you have to exert yourself don't forget to exhale. Exhaling as much as you can will help make your breathing more efficient.
- **F**ollow - Your doctor's recommendations may be the key to your breathing success. Be sure to take all the medications and follow all instructions your doctor has suggested to you. These recommendations will help maintain your overall well-being.
- **G**row – Follow the progress of your pregnancy and keep your regularly scheduled appointments to follow the growth of your baby.
- **H**elp – Don't be afraid to ask for help when a physical task is difficult or overwhelming. Seek help in understanding your pregnancy by using your available resources.
- **I**nhale - Make sure to inhale when you are involved in active motion. This allows for maximum amount of oxygenation of the blood that is being returned from your body.
- **J**ump - Jump to action! If you notice any changes in your condition alert your physician immediately.
- **K**eep calm - When you feel anxious or an episode of shortness of breath coming on stay calm and focus on your breathing.
- **L**ive efficiently - Live each day to its fullest!
- **M**ultivitamins – take your multivitamins.
- **N**ever miss a doctor's appointment; it can be vital for you and your baby's health.
- **O**rganize – your daily life.
- **P**riorities - Set your priorities. Use your energy on the things that matter most first and work down the list.
- **Q**uit smoking - If you smoke and currently have a breathing problem, you must quit smoking. Breathing efficiency and

quality starts improving from the time you put out your last cigarette.

- **R**emember – Remember your support system. This consist of your family, physicians, friends and anyone else that is involved in your well-being. If you need any help with your health, remember to contact the appropriate individual without delay.
- **S**leep – Get adequate sleep; remember your sleeping for two!
- **T**ake – It's important that you take all your medications as your doctor has prescribed them. Take advice. All of your healthcare providers are there to help you. It is also important NOT to stop any medication without consulting a health care provider.
- **U**nique – understanding the unique gift of pregnancy. This is a wonderful time in your life. Enjoy it.
- **V**entilation – It's important to keep the air in your house well ventilated and in constant circulation. Smoke, steam, dust, pets and other irritants may worsen your breathing. Ask your physician about proper ventilation methods.
- **W**ater – stay well hydrated, especially during periods of physical activity and warm temperatures.
- **eX**ercise – Slow, but steady, physical and strength training exercises can help you better your breathing efficiency. It may be difficult at first, but it will eventually help your breathing.
- **Y**esterday is gone! Focus on what you can do today to improve your health & well-being.
- **Z**est – Incorporate some activities in your life that make you happy. Being in a better mood can increase your overall health; help improve you breathing and outlook on life.

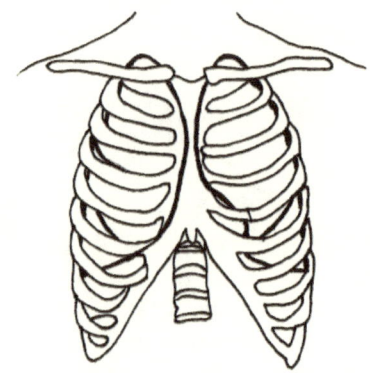

Blood tests that your physician may order

Complete Blood Count (CBC)

The CBC is a test that is ordered to get a profile of the blood cells in your body. The CBC provides information about the white blood cells (WBC), the red blood cell (RBC) and platelets that are in the blood. This information includes the number, type, size, shape, and some of the observations of the cells.

- White blood cells (WBC's) protect the body against infection
- Red blood cells (RBC's) carry oxygen
- Platelets help stop bleeding

Lipid Panel (Cholesterol Test)

The Cholesterol test or sometimes called the lipid test, is used to estimate your risk of developing heart disease. Cholesterol is important for your body to produce hormones and to aid with digestion. The fats you eat are stored in the liver and travel through the body along with cholesterol. Cholesterol particles are made up of proteins and fats that are bound together to form three main types of cholesterol. The Cholesterol test provides information on Low Density Lipoproteins (LDL), High Density Lipoproteins (HDL), Very Low Density Lipoproteins (VLDL), and Triglycerides. LDL is sticky and will stick to your arteries (Arteries are tubes that carries blood through your body, like pipes carry water to parts of your home). Target cholesterol goals are as follows: An LDL of <100 mg/dL, HDL of >40 mg/dL - and Triglycerides <150 mg/dL

- VLDL carries your triglycerides (Fats) to your fat cells.
- LDL is what remains after the fat has been delivered to its destination.
- HDL carries the remaining LDL back to the liver.
- Triglycerides are fats in your blood used to store energy when needed.

Glucose Testing

To correctly assess the sugar level in your body, a basic metabolic panel may be ordered to get an idea of a random glucose level. However, to better gauge how the sugar levels have been running in the body for the last 8-10 weeks, your physician may order a test, called the glycosylated hemoglobin (HemoglobinA1C). The A1C test is a blood test that measures how much sugar is attached to the hemoglobin protein that is present inside your red blood cells. It is important that your HgbA1C be checked routinely every 6 months and every 3-6 months for diabetics. The target glucose goal should be a glucose of <100 mg/dL and a hemoglobinA1C of 3.5%-5.5%.

Diabetes occurs when your body is unable to handle the amount of sugars in your blood. Insulin (A hormone made by the pancreas) moves sugars in your blood to your cells. Your cells use the sugars as fuel for energy. If you have too much sugar in your blood, your cells become non respondent to insulin's request to your cells to process sugar.

Comprehensive Metabolic Panel (CMP)

The CMP is routinely ordered as part of a blood work-up for a medical exam or your yearly physical. The test is usually conducted after you have been fasting. This test will test the electrolytes in your body as well as the liver and kidney function. The test may not tell you physician what exactly what is wrong with you but may give an idea of what may be causing the abnormal test results.

Kidney Tests

Blood Urea Nitrogen (BUN) and Creatinine

Urea is the waste product of your body after your body uses proteins. Urea is sent from your kidneys and cleared by your body as urine. Creatinine is a waste product in the formation of muscle. These tests are used to see how well your kidneys are removing waste in your body.

The estimated glomerular filtration rate (eGFR) is a calculation used to see how well your kidneys are working. Kidney disease is best treated when diagnosed in its early stages.

Urine Albumin

It may be that your physician orders this test. Urine albumin test checks urine for a protein called albumin. It is normally found in the blood and filtered by the kidneys. When the kidneys are working as they should, there may be a very small amount of albumin in the urine. But when the kidneys are damaged, abnormal, or under stress, they may cause albumin to leak. This finding is sometimes consistent with a finding called preeclampsia, which results due to uncontrolled blood pressure.

Oxytocin Testing

This is a test to see how your baby will do with the stress of contraction. Oxytocin is given via IV and when contractions occur fetal monitoring lets your physician know if the baby is ready for delivery. Please discuss this test with your physician.

Liver Tests

- ALP (alkaline phosphatase)
- ALT (alanine amino transferase, also called SGPT)
- AST (aspartate amino transferase, also called SGOT)
- Bilirubin

ALP, ALT and AST are enzymes found in the liver and other tissues. Bilirubin is a waste product when your liver eliminates old red blood cells. These enzymes are used to determine if your liver is working properly.

8 | Cardiac disease in Pregnancy

"A baby fills a place in your heart that you never knew was empty"

Unknown

Pregnancy causes many changes to your body which may add additional stress to your heart. The heart pumps blood around the body, and the blood carries oxygen and nourishment. If the pump does not work as well as normal, the developing baby may not get all the oxygen and food it needs. It may, therefore, not grow as well as normal (fetal growth restriction) or it may be born premature (or 'preterm' as we now say). With good neonatal care, many small babies can do well after they are born, but some may have a permanent handicap. For a few, this will be severe. You need to consider how you would cope with this if it happens. The cause of heart disease during pregnancy is unknown in most cases. It usually presents in late pregnancy, but it can occur up to 6 months after delivery. Heart disease should be considered in any pregnant woman who complains of increasing shortness of breath, especially those who lie flat at night. This chapter will focus on the diseases of the heart during pregnancy.

Ischemic Heart Disease

Pregnancy itself raises the risk of heart attack by three- to four-fold, with the risk being 30 times higher for women over the age of 40 years compared with women aged less than 20 years of age. Unfortunately, many of these risk factors are becoming increasingly common, and most women affected will not have many symptoms before pregnancy and no history of heart disease. The key component of discovering heart disease during pregnancy is seeing your physician immediately with any feeling of chest pain. All women with chest pain in pregnancy should have an electrocardiogram interpreted by someone who is skilled at detecting signs of cardiac disease.

Rheumatic Heart Disease

Rheumatic heart disease is often caused by a prior occurrence of strep throat. Many women affected by this condition will be unaware that they have valvular heart disease until they become symptomatic. This highlights the need for a particularly careful cardiovascular assessment at the beginning of pregnancy of all women. Mitral valve stenosis (the most common lesion and the one that carries the highest risk) is a difficult clinical diagnosis and clinical suspicion should warrant an echocardiography

Congenital heart disease

Congenital heart disease is a disease of the heart that you were born with. If you were born with a heart disease, you should consult your physician before pregnancy or as soon as you become pregnant. Congenital heart disease may cause shortness of breath, low oxygen levels in the blood or strain on your pulmonary system. If you have grown-up congenital heart disease (GUCH) and want to be (or are) pregnant, you should discuss with your obstetrician and cardiologist.

See your cardiologist

The cardiologist is the expert in heart disease, and their job is to keep you as healthy as possible. They will have complete knowledge of your condition, and they can explain to you the effect that pregnancy will have on your health. If they think pregnancy will be dangerous for you, they may advise you not to become pregnant. However, you should remember that ultimately this is a decision only you can make, in conjunction with your partner and with the knowledge of all the facts.

It is very important that full testing is carried out before pregnancy to establish how well your heart is working. This will enable the cardiologist to give you the most accurate advice, and the information gained will be vital in the proper care of a pregnancy. Pregnancy puts an increased strain on the heart. Sometimes, surgery to improve its function can be undertaken, which will make a subsequent pregnancy safer. The best advice is to see the obstetrician before you become pregnant

In addition, the tendency to have a heart defect is hereditary; if you have one, your baby will probably have a 3–5% risk (one in 20) of having one too (the risk varies somewhat, depending on the precise condition). This is about five times the average risk. If your partner has a heart problem, the risk is even higher. Nowadays, up to 80% of heart abnormalities can be detected using ultrasound scanning. This is usually offered to you between 11 and 24 weeks of the pregnancy (the later the scan, the bigger the baby, and the more detail can be seen). If an abnormality is detected, you will be offered the possibility of terminating the pregnancy. You will need to decide how you feel about this.

These days, much medical care, including antenatal care, is done as an 'outpatient'. However, if your heart has difficulty pumping well enough to meet both your needs and the needs of the developing baby, extra rest will be necessary. Sometimes, adequate rest can be obtained only by admitting the mother to hospital, where she needs to do nothing except grow the baby. In addition, close observation of your heart and of the developing baby may be necessary on a day-to-day basis. All this means that you need to plan for the possibility of spending quite a lot of time in hospital, and in a few cases this can be most of the pregnancy.

What will happen when I am pregnant?

The demand on the heart increases from very early pregnancy, as the hormones adjust the mother's body to help the developing baby (fetus). You should see your obstetrician very early (at about eight weeks from the beginning of the last period, which is about six weeks from conception of the baby). Your pregnancy should be jointly supervised by a cardiologist and an obstetrician, ideally at the same clinic. It is very important to see the obstetrician frequently, so that they can get used to you and how you are, and you can get to know them. This way, they will be much more able to pick up early signs of any problem developing. Depending on her cardiac status, the woman should be seen by an appropriately experienced consultant obstetrician every two to four weeks until 20 weeks, then every two weeks until 24 weeks, and then weekly thereafter.

At each visit, you will be asked about shortness of breath (especially at night) and your exercise tolerance (can you still climb stairs or walk at your normal pace), palpitations (irregular heart beat) and your own feelings of how things are going (for example, are you feeling the baby move). They will measure your pulse rate and rhythm, your blood pressure, whether you have any fluid collection at the ankles (edema), and the size of the uterus to judge how well the baby is growing. They will also listen to your lungs (again to check for any collection of fluid, or pulmonary edema) and your heart (to detect any changes in murmurs which might indicate deterioration in the functioning of a valve, or infection of the heart). You will also see a midwife who will advise you about the normal aspects of pregnancy and birth.

It is important to minimize the strain on the heart by vigorous treatment of any infections (for example chest, urinary). If the heart beat has any tendency to be irregular, drugs such as atenolol or digoxin may be given to control the rate. Regular scans to check on the growth of the baby will probably be necessary. If there is any anxiety about your condition, or that of your baby, you are likely to be admitted to hospital for rest and tests.

And finally...

Don't forget that if you decide to get pregnant, taking extra folic acid (easily obtainable from most pharmacies) for three months before and after conceiving will reduce substantially the risk of the baby having spina bifida (this applies to all women, not just those with heart disease).You should also make sure you have a good diet, and aim for a good body weight (not too fat or too thin). It is also advisable to get a blood test from your doctor to make sure that you are immune to rubella (German measles), because if you are not, it is a good idea to be vaccinated before you become pregnant (rubella is very dangerous to the baby if you become pregnant).And of course, if you are a smoker, you should do your very best to stop before you become pregnant.

Antepartum Care (Before Pregnancy)

Because there are so many types of cardiac disease, often with very different implications, it is important that a risk assessment of any woman with a heart murmur or a history of any cardiac defect should be carried out early in pregnancy by an obstetrician, cardiologist, and anesthesiologist. Women at low risk can be identified and returned to routine care. Women at significant risk of adverse events during pregnancy should be seen regularly in the antenatal clinic, whenever possible by the same consultant obstetrician, who should have appropriate competencies in this field. Cardiovascular assessment should be carried out at every antenatal clinic. Blood pressure should be measured manually with a sphygmomanometer. Auscultation to assess any change in murmur or any lung changes associated with pulmonary edema is recommended in all cases of significant cardiac compromise (which will have been identified early in pregnancy at the joint clinic). Women with cyanotic heart disease should have their oxygen saturations checked periodically (each trimester or more often if there are any clinical signs of deterioration).

All women with structural congenital heart disease should be offered a fetal echocardiogram during the second trimester to be carried out by an accredited pediatric/fetal cardiologist.

A further multidisciplinary meeting should take place at 32–34 weeks of gestation to establish a plan of management for delivery. Important features of such a plan include deciding who should be involved in supervising the labor, whether a caesarean section is appropriate, whether bearing down is advisable in the second stage and appropriate prophylaxis against postpartum hemorrhage. The plan should also include postpartum management, including whether prophylaxis against thrombosis is appropriate, the length of postpartum stay in hospital, and the timing of cardiac and obstetric review.

Intrapartum Care (During Pregnancy)

The general principle of intrapartum management is to minimize cardiovascular stress. In most cases, this will be achieved by the use of early slow incremental epidural anesthesia and assisted vaginal delivery. Caesarean section is usually necessary only for obstetric indications.

Some women will benefit from specialist care at tertiary units. The decision about the optimum place for antenatal and intrapartum care should be made in conjunction with obstetricians and cardiologists at tertiary units, known to specialize in the management of women with heart disease in pregnancy. Appropriate tertiary units will have high-dependency and intensive care units suitable for the care of pregnant women with significant heart disease.

Postpartum (After Delivery)

The length of recommended stay in hospital and any suggested special measures (such as anticoagulation, or observation in a high-dependency area) should be given by your treating physician.

Preeclampsia

Preeclampsia is a pregnancy complication characterized by high blood pressure and signs of damage to another organ system, most often the liver and kidneys. Preeclampsia usually begins after 20 weeks of pregnancy in women whose blood pressure had been normal.

Left untreated, preeclampsia can lead to serious — even fatal — complications for the mother as well as the baby. If you have preeclampsia, the only cure is delivery of your baby.

If you're diagnosed with preeclampsia too early in your pregnancy to deliver your baby, you and your doctor face a challenging task. Your baby needs more time to mature, but you need to avoid putting yourself or your baby at risk of serious complications.

Symptoms of Preeclampsia

Preeclampsia sometimes develops without any symptoms. High blood pressure may develop slowly, or it may have a sudden onset. Monitoring your blood pressure is an important part of prenatal care because the first sign of preeclampsia is commonly a rise in blood pressure. Blood pressure that exceeds 140/90 millimeters of mercury (mm Hg) or greater — documented on two occasions, at least four hours apart — is abnormal.

Other signs and symptoms of preeclampsia may include excess protein in your urine (proteinuria) or additional signs of kidney problems; severe headaches; changes in vision, including temporary loss of vision, blurred vision or light sensitivity; upper abdominal pain, usually under your ribs on the right side; nausea or vomiting; decreased urine output; decreased levels of platelets in your blood (thrombocytopenia); impaired liver function; and shortness of breath, caused by fluid in your lungs.

9 | Pregnancy & Other Health Problems

"The good physician treats the disease; the great physician treats the patient who has the disease."

Sir William Osler (1849-1919)

During the course of your pregnancy, you may be faced with other medical problems as well. This chapter will outline the other conditions you might face.

Sleep Disorders

The human body needs sleep for recovery and restitution. It is an active state essential for mental and physical restoration. Regular inability to get a good night's sleep may be indicative of a sleep disorder.

Obstructive Sleep Apnea (OSA)

Obstructive Sleep Apnea is the most common sleep disorder. It is a serious and potentially life-threatening condition that often goes undiagnosed. Loud snoring may signal that something is wrong with breathing during sleep and reflect presence of OSA. The condition affects at least 2-4% of middle-aged adults. Approximately 95% of the affected population remains undiagnosed and untreated.

Some of the warning signs of OSA are, but not limited to excessive daytime fatigue and sleepiness, snoring, falling asleep at inappropriate times, poor performance at home or at work, and cessation of breathing at night.

What are the Consequences of Untreated OSA?

When OSA goes untreated it may lead to elevations in blood pressure, heart arrhythmias or even heart failure, poor oxygenation, heart attack, cardiovascular accidents such as stroke. Excessive daytime sleepiness and fatigue may lead to traffic and industrial accidents.

How can Obstructive Sleep Apnea be Treated?

Currently there are a few options for the treatment of OSA. The most effective option is a continuous positive air pressure (CPAP) machine which uses air to help keep your airways open while you sleep. Those who are intolerant to the CPAP device may be considered for surgery or an oro-dental device. The most conservative treatment option is weight loss in conjunction with lifestyle modification. You are encouraged to discuss these therapeutic options with your physician if you suspect you may have OSA symptoms or you have been diagnosed to have OSA.

Sleep Hygiene

Poor sleep habits (referred to as hygiene) are among the most common problems encountered in our society. We stay up too late and get up too early. Below are some essentials of good sleep habits. Many of these points will seem like common sense. But it is surprising how many of these important points are ignored by many of us.

- Fix a bedtime and an awakening time. Do not be one of those people who allow bedtime and awakening time to drift. The body "gets used" to falling asleep at a certain time, but only if this is relatively fixed. Even if you are retired or not working, this is an essential component of good sleeping habits.
- Avoid napping during the day. If you nap throughout the day, it is no wonder that you will not be able to sleep at night. The late afternoon for most people is a "sleepy time". Many people will take a nap at that time. This is generally not a bad thing to do, provided you limit the nap to 30-40 minutes and can sleep well at night.
- Avoid alcohol.
- Avoid caffeine 4-6 hours before bedtime. This includes caffeinated beverages such as coffee, tea and many sodas, as well

as chocolate, so be careful. Avoid heavy, spicy, or sugary foods 4-6 hours before bedtime. These can affect your ability to stay asleep. Exercise regularly, but not right before you go to bed. Regular exercise, particularly in the afternoon, can help deepen sleep. Strenuous exercise within the 2 hours before bedtime, however, can decrease your ability to fall asleep.

- Exercise a digital curfew. Ban digital devices from the bedroom. Technology can alienate people. As smart phones continue to burrow their way into our lives, wearable devices such as Google Glass® threaten to invade out personal space even further. Draw a gadget free line in the bedroom. Don't sleep with your cellphone, leave it downstairs or in the living room overnight.

Your Sleeping Environment

- Use comfortable bedding. Uncomfortable bedding can prevent good sleep. Evaluate whether or not this is a source of your problem and make appropriate changes.
- Find a comfortable temperature setting for sleeping and keep the room well ventilated. If your bedroom is too cold or too hot, it can keep you awake. A cool (not cold) bedroom is often the most conducive to sleep.
- Reserve the bed for sleep. Don't use the bed as an office, workroom or recreation room. Let your body know that the bed is associated with sleeping.

Getting Ready For Bed

- **Eat Light:** Try a light snack before bed. Warm milk and foods high in the amino acid tryptophan, such as bananas, may help you to sleep.
- **Develop a Routine:** Practice relaxation techniques before bed. Relaxation techniques such as yoga, deep breathing, and other may help relieve anxiety and reduce muscle tension.
- **Keep a To Do List:** Don't take your worries to bed. Leave your worries about job, school, daily life, etc., behind when you go to bed. Some people find it useful to assign a "worry period" during the evening or late afternoon to deal with these issues.
- **Establish a Pre-sleep Ritual:** Pre-sleep rituals, such as a warm bath or a few minutes of reading, can help you sleep.

- **Find a Sleeping Position:** If you don't fall asleep within 15-30 minutes, get up, go into another room and read until sleepy.
- **Power Down:** Any device with a screen (TV, tablet PC, laptop, iPad) emits a blue spectrum light that can inhibit the brain production of melatonin that induces sleep. Some people find that the radio helps them go to sleep. Since radio is a less engaging medium than TV, this might be a better idea.

High blood pressure (Hypertension)

Blood pressure is the force of blood pushing against the inside of your blood vessels, called arteries, as your heart pumps. High blood pressure is a serious condition that causes your heart to work harder. It is also called hypertension. It can cause heart disease, stroke, kidney failure, blood vessel disease and may be harmful to the baby (preeclampsia/eclampsia). Those who are more likely to have high blood pressure include:

- African Americans
- Men over the age of 45 and women over the age of 55
- People with a family history of high blood pressure
- Women who are pregnant, or who take birth control pills, or hormone replacement therapy
- People with health conditions like thyroid disease, chronic kidney disease, or sleep apnea
- Those who take certain medicines, such as asthma medicines and cold-relief products

Your chances of having high blood pressure increase if you:

- Are overweight
- Eat foods high in salt
- Do not get regular exercise
- Smoke
- Drink alcohol heavily

There is no way to tell that you have high blood pressure. The only way to know if you have high blood pressure is to have it checked. The following are some key points regarding your blood pressure measurement:

- There are two blood pressure numbers. Systolic (top number) - the pressure when your heart pumps the blood out of your body.
- Diastolic (bottom number) - the pressure when your heart is resting in between beats.
- Your blood pressure should be less than 120/80 mmHg. (120 is your systolic number, 80 is your diastolic number)
- High blood pressure is when your blood pressure is 140/90 mmHg or greater.
- "Pre-hypertension" is when your blood pressure is greater than 120/ 80 mmHg, but less than 140/90 mmHg. When you have pre-hypertension, you may be at risk for high blood pressure and other health related problems.
- If you have diabetes or kidney problems, your blood pressure should be less than 130/80 mmHg.

What changes can I make in my life if I have high blood pressure?

High blood pressure needs to be controlled. You can change or control some lifestyle habits that will help treat, prevent or delay high blood pressure. These include:

- healthy eating - choosing a low salt or no salt diet
- staying physically active
- keeping or getting to a healthy weight
- quitting smoking
- avoid alcohol
- dealing with stress in a healthy way
- taking your high blood pressure medicine as prescribed
- keeping all appointments with your doctor

Diabetes and Pregnancy

When you eat, your body breaks down sugar and starches from food into glucose to use for energy. Your pancreas (an organ behind your stomach) makes a hormone called insulin that helps your body keep the right amount of glucose in your blood. When you have diabetes, your body doesn't make enough insulin or can't use insulin well, so you end up with too much sugar in your blood. This can cause serious health problems, like heart disease, kidney failure, and blindness. It's really important to get treatment for diabetes to help prevent problems like these.

Gestational diabetes is a kind of diabetes that can happen during pregnancy. If untreated, gestational diabetes can cause problems for your baby, like premature birth and stillbirth. Gestational diabetes usually goes away after you have your baby; but if you have it, you're more likely to develop diabetes later in life. Most pregnant women get a test for gestational diabetes at 24 to 28 weeks of pregnancy.

Gestational Diabetes and Pregnancy

Gestational diabetes can be controlled and treated during pregnancy. But if not treated, it can cause problems during pregnancy, including:

- Preeclampsia. This is when a pregnant woman has high blood pressure and signs that some of her organs, like her kidneys and liver, may not be working properly. Signs of preeclampsia include having protein in the urine, changes in vision and severe headaches.

- Premature birth. This is birth before 37 weeks of pregnancy. Premature babies are more likely than full-term babies to have health problems at birth and later in life.

- Having a very large baby, weighing more than 9 pounds. Weighing this much makes your baby more likely to get hurt during labor and birth. Large babies are more likely to be obese or have diabetes later in life.

- Stillbirth. This is when a baby dies in the womb after 20 weeks of pregnancy.

Risk of Developing Gestational Diabetes increases if you have/are:

- Family history of diabetes
- Overweight before pregnancy
- Previous birth of baby of 9 pounds or stillbirth
- Had GDM with previous pregnancy

- Age over 25 years old
- Ethnic groups: Hispanic, African American, Asian, Native American, Pacific Islander
- Chronic use of medications that increase blood glucose levels: steroids, betamimetics, or atypical antipsychotics

The American Diabetes Association recommends lifestyle management of gestational diabetes before medications are introduced. Hemoglobin A1C should be maintained at 6% or less without hypoglycemia. In general, insulin is preferred over oral agents for treatment of gestational diabetes.

There tends to be a spike in insulin resistance in the second or third trimester; women with preconception diabetes, for example, may require frequent increases in daily insulin dose to maintain glycemic levels, compared to the first trimester. A baseline ophthalmology exam should be performed in the first trimester for patients with preconception diabetes, with additional monitoring as needed.

Following pregnancy, screening should be conducted for diabetes or prediabetes at six to 12 weeks' postpartum and every one to three years afterward. Women with gestational diabetes should maintain lifestyle changes, including diet and exercise, to reduce the risk for Type 2 Diabetes Mellitus later in life.

Thyroid Disease

The thyroid is located in the neck, in front of the windpipe. The hormones produced by the thyroid gland influence your heart rate, your metabolism, and many other aspects of your health.

Sometimes, the thyroid gland produces too much or too little of the thyroid hormone (thyroxine) that keeps the body functioning normally.

- Hyperthyroidism is the disorder that occurs if the thyroid gland is too active.
- Hypothyroidism is the disorder that occurs if the thyroid gland isn't active enough.

Some women have a thyroid disorder that began before pregnancy. Others develop thyroid problems for the first time during pregnancy or soon after delivery.

An untreated thyroid disorder during pregnancy is a danger to both mother and baby. For mothers, the risks include preeclampsia. For babies, the risks include preterm birth, decreased mental abilities, thyroid disorder, and even death. But with proper treatment, most women with thyroid disorders can have a healthy baby.

Signs and symptoms of hyperthyroidism:

- Nervousness, anxiety attacks, or irritability
- Sudden weight loss
- Rapid heartbeat, irregular heartbeat, or pounding of the heart (palpitations)
- Shaking hands and fingers
- Inappropriate sweating
- Increased sensitivity to heat
- More frequent bowel movements
- Changes in menstrual patterns
- Fatigue
- Muscle weakness
- Difficulty sleeping

Most medications used to treat hyperthyroidism are safe in pregnancy. But you must consult with your physician.

Signs and symptoms of hypothyroidism include:

- Unexplained weight gain
- Constipation
- Increased sensitivity to cold
- Dry skin
- Hoarseness
- Heavier than normal menstrual periods
- Depression
- Muscle and joint aches

- Muscle weakness
- Hypothyroidism is treated with a medication called Synthroid.

Routine blood tests are required during pregnancy to adjust the doses of your medication for both Hyperthyroidism and Hypothyroidism.

Kidneys and Lung Disease

The kidneys are organs in your body that have a lot of important function. These include getting rid of all waste products via the making of urine. The kidneys are powerful set of organs that perform the following functions:

- Remove waste products from the body
- Remove drugs and toxins from the body
- Balances the electrolytes in your body
- Regulates blood pressure
- Regulates Vitamin D production
- Regulates your hemoglobin count and prevents anemia

The kidneys work hand in hand with your heart and lungs to regulate the body's function. In the setting of pregnancy, the kidneys may be starved of necessary oxygen. When the lungs are not able to properly oxygenate the blood, it's called hypoxemia, which will ultimately cause the kidney to not function properly.

What happens when kidneys don't work?

- You are unable to get rid of your waste products and toxins build up and as a result you feel fatigue, weakness, tremors, decreased urine output are symptoms of kidney disease.
- You are unable to regulate blood pressure and fluid, resulting in hypertension or elevated blood pressure.
- If you cannot make the hormone for red blood cells, you get anemic or your hemoglobin amount drops.
- If you cannot make vitamin D, it leads to bone loss.
- You get puffiness around your eyes, hands and feet
- Your blood pressure on physical exam can be elevated.

How is Kidney Disease Detected?

Early detection and treatment of chronic kidney disease are extremely important for prevention of kidney problems.   3 simple tests can help tremendously.

- A test for protein and blood in the urine. This can be obtained by giving a urine sample to your doctor and getting a urinalysis done. When there is injury to the filtering unit of the kidney, the kidneys start losing protein and blood in the urine and you can only notice this microscopically.
- A test for blood creatinine or blood urea nitrogen. Your doctor should use your results, along with your age, race, gender and other factors, to calculate your glomerular filtration rate (GFR). Your GFR tells how much kidney function you have.
- An elevated blood pressure can be an earlier sign of kidney disease.

10 | Pregnancy and Tobacco

> *"Smoking is hateful to the nose,
> harmful to the brain and dangerous to
> the lungs."*
>
> *King James I (1566-1625)*

By now, you are well aware that smoking damages your lungs. If you smoke, quitting is the number one thing that you can do to improve your standard of living, more than diet, exercise, medications, or rehab. It is important to note that no matter how severe your lung disease is, smoking decreases quality of life. If you do not smoke, but someone in your household does, even if they don't smoke inside, you may be at increased risk of smoke-related problems. Even today, Tobacco use is the second cause of death globally (after high blood pressure) and is currently responsible for killing one in ten adults worldwide. Below are facts about smoking, which illustrate just how important it is for someone with any stage of disease to kick the habit.

What does smoking do to my lungs?

Damages the airways

- Your airways will become inflamed.
- The little hair like structures, called cilia, that usually move back and forth to sweep particles out of the airways will stop working normally. Cilia in the airways move particles out of the airways, and are paralyzed by tobacco smoke.
- Your large airways will produce more mucus, which can cause a chronic cough. This is called chronic bronchitis and is part of chronic obstructive pulmonary disease (COPD). You will cough and produce phlegm most of the time.

Smoking during Pregnancy

Smoking during pregnancy is bad for you and your baby. Tobacco is a plant whose leaves are used to make cigarettes and cigars. Tobacco contains a drug called nicotine. Nicotine is what makes you become addicted to smoking. When you smoke during pregnancy, chemicals like nicotine, carbon monoxide, and tar pass through the placenta and umbilical cord into your baby's bloodstream.

These chemicals are harmful to your baby. They can lessen the amount of oxygen that your baby gets. This can slow your baby's growth before birth (low birth weight) and can damage your baby's heart, lungs, and brain.

Quitting smoking, even if you're already pregnant, can make a big difference in your baby's life.

Smoking harms nearly every organ in the body and can cause serious health conditions, including cancer, heart disease, stroke, gum disease, and eye diseases that can lead to blindness.

How can smoking affect your pregnancy?

If you smoke during pregnancy, you're more likely than nonsmokers to have:

- Preterm labor. This is labor than starts too early, before 37 weeks of pregnancy. Preterm labor can lead to premature birth.
- Ectopic pregnancy. This is when a fertilized egg implants itself outside of the uterus (womb) and begins to grow. An ectopic pregnancy cannot result in the birth of a baby. It can cause serious, dangerous problems for the pregnant woman.
- Bleeding from the vagina
- Problems with the placenta, like placental abruption and placenta previa. The placenta grows in your uterus (womb) and supplies the baby with food and oxygen through the umbilical cord. Placental abruption is a serious condition in which the placenta separates from

the wall of the uterus before birth. Placenta previa is when the placenta lies very low in the uterus and covers all or part of the cervix. The cervix is the opening to the uterus that sits at the top of the vagina.

- Baby having birth defects, including birth defects in the mouth called cleft lip or cleft palate. Birth defects are health conditions that are present at birth. They change the shape or function of one or more parts of the body. They can cause problems in overall health, in how the body develops or in how the body works.

- If you smoke during pregnancy, you're more likely to have a miscarriage or a stillbirth.

- Your baby may experience sudden infant death syndrome (also called SIDS). This is the unexplained death of a baby younger than 1 year old.

What is secondhand smoke?

Secondhand smoke is smoke you breathe in from someone else's cigarette, cigar, or pipe. Being around secondhand smoke during pregnancy can cause your baby to be born with low birthweight.

Secondhand smoke also is dangerous to your baby after birth. Babies who are around secondhand smoke are more likely than babies who aren't to have health problems, like pneumonia, ear infections and breathing problems, like asthma, bronchitis and lung problems. They're also more likely to die of SIDS.

Can you just cut down on smoking? Or do you have to quit?

If you smoke, you may think that light or mild cigarettes are safer choices during pregnancy. This is not true. Or you may want to cut down rather than quit smoking altogether. It's true that the less you smoke, the better for your baby. But quitting is best.

The sooner you quit smoking during pregnancy, the healthier you and your baby can be. It's best to quit smoking before getting pregnant. But quitting any time during pregnancy can have a positive effect on your baby's life.

Benefits of smoking cessation in general

Time period	Result
Within 20 minutes	Blood pressure drops to near that of before the last cigarette. Temperature of hands and feet increases to normal.
Within 12 hours	Carbon monoxide level drops to normal
Within 24 hours	The risk of myocardial infection decreases
Within 2-3 weeks	Circulation improves and lung function increases
Within 1-9 months	Coughing, sinus congestion, fatigue, and shortness of breath decrease
Within 1 year	The excess risk of heart disease is half that of a smoker's
Within 5 years	The risk of stroke reduces to that of a nonsmoker's
Within 10 years	The risk of many cancers decreases, including lung, mouth, and throat cancer
Within 15 years:	The risk of heart disease reduces to that of a nonsmoker's

Specific benefits of smoking cessation for pregnant patients

- After you stop smoking, more nutrition will go to your baby to help him/her grow.
- After you stop smoking, your chances of having a healthy baby increase, and the baby is more likely to have a healthy childhood.

- After you stop smoking, you will have more energy and may feel less stressed.

- After you stop smoking, you'll breathe easier and you will be better able to keep up with your active, healthy baby.

- After you stop smoking, you'll reduce your risk for cancer, cardiovascular, and other diseases so you can be around a long time to be a good mother.

*Adapted with permission from You and your baby, American Lung Association: www.lungusa.org

---

How can you quit smoking?

Select a Quit Date

No one pretends giving up smoking is easy, but if you have made up your mind to quit, you can succeed. It is important to set a date on which you plan to quit smoking (quit-date) and mentally prepare to achieve you set out goal. Use simple tricks to reduce your urge to smoke and help you quit. Look for triggers and plan to avoid them. If you need assistance to quit smoking, please consult your physician or call 1-800 QUITNOW. Do not use any nicotine replacement products before speaking to a medical professional.

What are some tips to help you quit smoking?

Try these tips to help you quit smoking:

- Write down your reasons for quitting. Look at the list when you think about smoking.
- Choose a quit day. On this day, throw away all your cigarettes or cigars, lighters and ashtrays.
- Ask your partner or a friend to help you quit. Call that person when you feel like smoking. Stay away from places, activities or people that make you feel like smoking.

- Keep yourself busy. Go for a walk to help keep your mind off smoking. Use a small stress ball or try some needlework to keep your hands busy. Snack on veggies or chew gum to keep something in your mouth.
- Discuss your quitting options with your healthcare provider.
- Look for programs in your community or where you work that can help you stop smoking. These are called smoking cessation programs. Ask your health care provider about programs in your area. Ask your employer to see what services are covered by health insurance.

Don't feel badly if you can't quit right away. Keep trying! You're doing what's best for you and your baby.

How to Prevent Relapses of Smoking

- Remind yourself why you gave up smoking in the first place.
- Move away from the area of smokers.
- Keep busy to distract your mind: daily exercise is a good 'distraction' to promote continued abstinence, while counteracting weight gain.
- Drink plenty of water and take deep breaths.

Beware

Some triggers for smoking only reveal themselves after you try to live without cigarettes. Tricks that work for some people may not work for others, so quitting can involve trial and error. Keep going! Ask your doctor or nurse for help. The most important thing is to be determined and persistent.

If at first you don't succeed, try again...

Nicotine addiction is very powerful and only 5–10% of 'quit attempts' are successful. Withdrawal symptoms, such as craving, irritability, inability to sleep, mood swings, hunger and headache, that occur when the brain is looking for a new fix of nicotine, are a common reason for relapsing and treatment can help this.

Is it safe to use e-cigarettes during pregnancy?

Electronic cigarettes (also called e-cigarettes or e-cigs) look like regular cigarettes. But instead of lighting them, they run on batteries. E-cigarettes contain liquid that includes nicotine, flavors (like cherry or bubble gum) and other chemicals. When you use an e-cigarette, you puff on a mouthpiece to heat up the liquid and create a mist (also called vapor) that you inhale. Using an e-cigarette is called vaping. More research is needed to better understand how e-cigarettes may affect women and babies during pregnancy. If you're pregnant and using e-cigarettes or thinking about using them, please discuss with your health care provider.

Smoking & Pregnancy

Smoking can cause problems for a woman trying
to become pregnant or who is already pregnant,
and for her baby before and after birth.

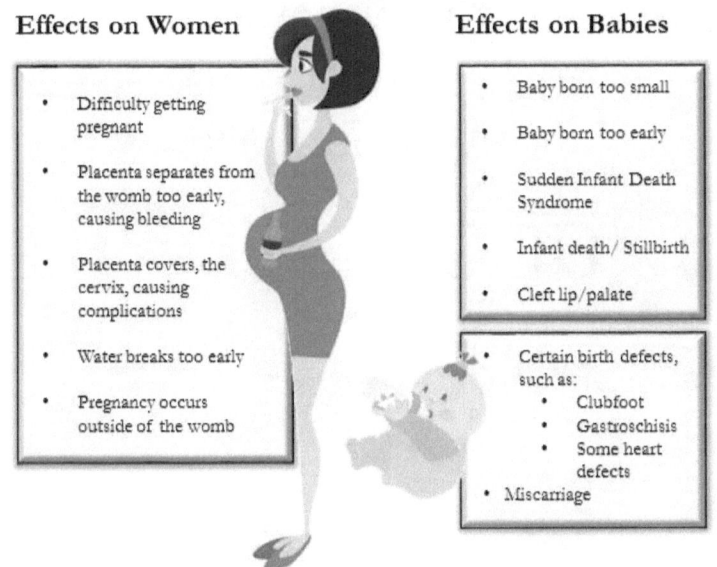

Effects on Women

- Difficulty getting pregnant

- Placenta separates from the womb too early, causing bleeding

- Placenta covers, the cervix, causing complications

- Water breaks too early

- Pregnancy occurs outside of the womb

Effects on Babies

- Baby born too small

- Baby born too early

- Sudden Infant Death Syndrome

- Infant death/ Stillbirth

- Cleft lip/palate

- Certain birth defects, such as:
  - Clubfoot
  - Gastroschisis
  - Some heart defects
- Miscarriage

"

The easiest way to
let go of a

BAD HABIT

Is to give your time
and attention to
developing a

GOOD ONE!

"

11 | Diet & Nutrition

"Eating healthy and in the right amounts can have wonderful effects on the prenatal development of your child."

Unknown

Proper nutrition contributes to overall wellness and is essential for individuals with chronic lung disease. A healthy body is better able to fight off infections, thus preventing simple colds from progressing into a more serious lung infection. If illness does occur, a well-nourished body helps to produce a better response to treatment and therefore helps you get better faster. You may consult a nutritionist, especially if you are dealing with additional problems such as heart disease or diabetes. The discussion in this chapter is general but please consult your physician for your individual needs.

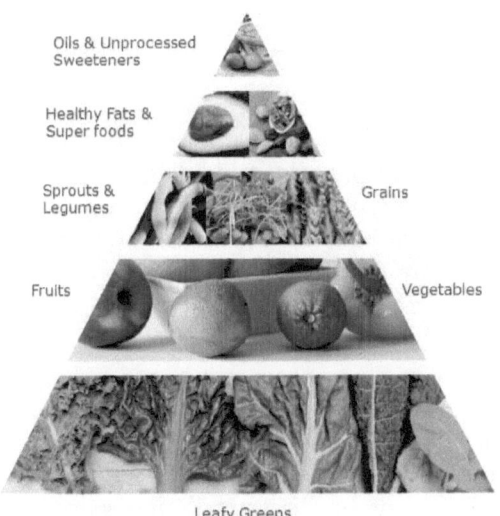

A healthy person should aim to get 45% to 65% of their calories from carbohydrates, with active individuals aiming for 55% to 65%. As for protein, a healthy individual should aim to get about 10% to 35% of their calories from proteins. Fats are also essential to our well-being, and as such, a healthy individual should aim to get 20% to 35% of their calories from fat.

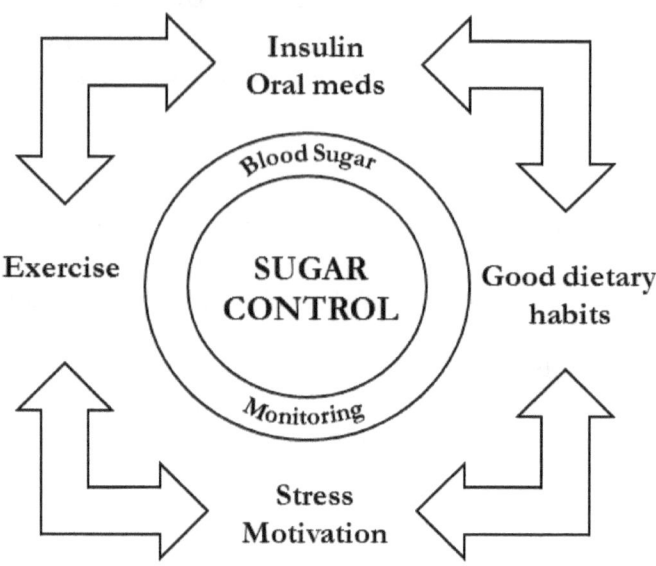

Proteins

While protein can be used for energy when carbohydrates and fat are in short supply, protein's major role is building muscle, making blood and other body tissues. Many people consume meat products as a source of protein. It is important to know red meats, such as beef, are linked to increased risk of heart disease. Good sources of protein include eggs, poultry (white meat), soy, and whey.

Carbohydrates

Carbohydrates are the body's main source of energy. They may come in the form of simple sugars, such as sucrose and fructose, or in wheat products, such as breads and pastas. Eating extra carbohydrate gives runners energy, but it will give patients with history of lung disease carbon dioxide.

Fats

Fats are a common component of many foods. It is important to keep in mind that not all fats are bad and they might not be associated with cholesterol. Foods that are high in cholesterol and saturated fats come mostly from animal sources. To differentiate between saturated and unsaturated fats, you can see how they act at certain temperatures. Saturated fats turn solid in cool temperatures, an example being butter or the layer of fat on top of a pot of chicken soup that's been in the refrigerator. Polyunsaturated fats do not contain cholesterol. These are fats are from plant sources and they remain liquid at cold temperatures, for example olive oil.

When you consume fat, make an effort that it is the polyunsaturated kind. Avoid animal fats such as butter and cut down on fatty meats, as these foods are high in cholesterol. Before you buy prepared products, read the labels. Many prepared foods list the cholesterol content on the ingredient panel. If you can buy either a product made with butter or one made with corn oil, choose the one made with corn oil.

More recently, as a community, we have become of aware of a different type of fat called a trans-fats. Trans-fats are a type of unsaturated fat that have a different chemical configuration than typical saturated and unsaturated fats. These trans-fats are created by the processing and hydrogenation of unsaturated fats. These types of fats may be found naturally in the plant kingdom and in certain meat products like beef. Trans-fats have been associated with increased incidence of cardiac disease. They should be avoided whenever possible.

Healthy vs. Less Healthy Fats

Healthy Fats Monounsaturated Fats	Less Healthy or Unhealthy Fats Saturated Fats
Olive oil	Butter
Peanut oil	Lard
Sesame oil	Tropical oils such as coconut, palm, palm kernel, cocoa butter
Variety of nut and seed oils: peanut, almond, macadamia nut, sesame	Fatty meats
Avocados, olives	Whole milk
Polyunsaturated Fats	**Trans (Hydrogenated) Fats**
Corn oil	Shortening
Soybean oil	Hard stick margarine (check labels)
Oil found in fish, salmon, tuna, mackerel, sardines, herring, anchovies	
Safflower oil	Many fried foods, such as French fries, chicken nuggets, fish sticks
Variety of nut and seed oils: walnut, pumpkin, flax	
* Heart-healthy spreads, such as Benecol and Take Control (if used two to three times a day in place of regular margarine or butter, these products may lower cholesterol by up to 14 percent	

Portion Control

Portion control is an important part of any diet and weight management plan. Our stomachs have adapted to our diet and lifestyle over many years. When you eat a really large meal, the diaphragm cannot move as far down, so the lungs do not fill as well. Instead of eating three big meals a day, try dividing your day's food into five or six smaller portions. This way, your stomach will not fill as much after each meal. You can eat a smaller breakfast, lunch, and dinner and supply the rest of your nutritional needs for the day by having two or three small snacks.

Instead of ...	Try ...
Potato chips	Soy chips
A bag of M&M's	Trail mix (with a few M&M's)
Before-dinner pretzels	Before-dinner edamame
Fried chicken	Grilled chicken
Hot fudge sundae	Frozen yogurt with fruit and granola
Taco chips and cheese sauce	Veggies and cheese sauce
French fries	Roasted sweet potato chips
Anything on white bread	Anything on whole wheat
A soft drink	A fruit smoothie
Sugar cookies	Whole-grain Fig Newtons

Nutrition in Pregnancy

Nutrition in pregnancy can affect maternal health as well as infant size and well-being. Pregnant women should have nutrition counseling early in prenatal care and access to supplementary food programs if necessary. Counseling should stress abstention from alcohol, smoking, and recreational drugs. Caffeine and artificial sweeteners should be used only in small amounts.

Recommendations regarding weight gain in pregnancy should be based on maternal body mass index (BMI) pre-conceptionally or at the first prenatal visit. According to the Institute of Medicine guidelines, total weight gain should be 25-35 lbs (11.3-15.9 kg) for normal weight women (BMI of 18.5-24.9) and 15-25 lbs (6.8-11.3 kg) for overweight women. For obese women (BMI of 30 or greater), weight gain should be limited to 11-20 lbs

(5.0-9.1 kg). Excessive maternal weight gain has been associated with increased birth weight, as well as postpartum retention of weight. Not gaining weight in pregnancy, conversely, has been associated with low birth weight. Nutrition counseling must be tailored to the individual patient.

The increased calcium needs of pregnancy (1200 mg/ day) can be met with milk, milk products, green vegetables, soybean products, corn tortillas, and calcium carbonate supplements.

The increased need for iron and folic acid should be met with foods as well as vitamin and mineral supplements. (See Anemia section.) Megavitamins should not be taken in pregnancy, as they may result in fetal malformation or disturbed metabolism. However, a balanced prenatal supplement containing 30-60 mg of elemental iron, 0.5-0.8 mg of folate, and the recommended daily allowances of various supplements can decrease the risk of neural tube defects in the fetus. For this reason, the United States Public Health Service recommends the consumption of 0.4 mg of folic acid per day for all pregnant and reproductive age women. Women with a prior pregnancy complicated by neural tube defect may require higher supplemental doses as determined by their providers.

Vitamin Supplements

Vitamin supplementation is an important part of your pregnancy. All nutrients are important, but these six play a key role in your baby's growth and development during pregnancy: Folic acid, Iron, Calcium, Vitamin D, Omega-3 Fatty Acids and Iodine. Please discuss with your physician the need for these vitamins and supplements.

Nutrition Pearls

- Prenatal vitamins with iron and folic acid should be consumed.
- Supplements that are not specified for pregnant women should be avoided as they may contain dangerous amounts of certain vitamins.
- Caffeine intake should be decreased to 0-1 cup of coffee, tea, or caffeinated cola daily.
- Avoid eating raw or rare meat as well as fish known to contain elevated levels of mercury.

Food Factors that Modify Absorption of Dietary Iron

Inhibitors of absorption of dietary non-heme iron	Example of food source	Enhancers of absorption of dietary non-heme iron	Example of food source
Tannins	Tea	Vitamin C	Citrus fruit Strawberries Kiwifruit Capsicum Tomatoes Broccoli
Oxalates	Spinach	Meat proteins	
Phytates	Cereals Legumes Nuts		
Polyphenols	Tea Coffee Red wine		
Calcium*	Dairy foods		

*Also inhibits absorption of heme iron

"

Sometimes when things are falling apart they may actually be falling

Into place

12 | Traveling During Pregnancy

"Traveling gives you some perspective of what the rest of the world is like. I think that having the courage to step out of the norm is the most important thing."

Meghan Markle (b. 1981)

As with any other aspect of being pregnant, planning ahead and taking precaution can allow you to travel with a free mind. It's good to get out of the house and enjoy life. There are a few pregnancy travel related concerns; this chapter is provided to help make your travels the safest and most comfortable it can be.

As long as there are no identified complications or concerns with your pregnancy, it is generally safe to travel at all times during your pregnancy. The ideal time to travel during pregnancy is the second trimester. In most cases, you are past the morning sickness of the first trimester and several weeks from the third stage of pregnancy, when you are more easily fatigued.

Traveling During Pregnancy

Whether you are going by car, bus, or train, it is generally safe to travel while you are pregnant; however, there are some things to consider that could make your trip safer and more comfortable.

- It is essential to buckle-up every time you ride in a car. Make sure that you use both the lap and shoulder belts for the best protection of you and your baby.

- Keep the air bags turned on. The safety benefits of the air bag outweigh any potential risk to you and your baby.
- Buses tend to have narrow aisles and small restrooms. This mode of transportation can be more challenging. The safest thing is to remain seated while the bus is moving. If you must use the restroom, make sure to hold on to the rail or seats to keep your balance.
- Trains usually have more room to navigate and walk. The restrooms are usually small. It is essential to hold on to rails or seat backs while the train is moving.
- Try to limit the amount of time you are cooped up in the car, bus, or train. Keep travel time around five to six hours.
- Use rest stops to take short walks and to do stretches to keep the blood circulating.

Traveling By Air During Pregnancy

Traveling by air is considered safe for women while they are pregnant. In fact, most airlines allow pregnant women to travel through their eighth month. The International Air Travel Association recommends that expecting mothers in an uncomplicated pregnancy should avoid travel from the 37th week of pregnancy through birth. Below are some tips that may be useful while travelling by air.

Most airlines have narrow aisles and smaller bathrooms, which makes it more challenging to walk and more uncomfortable when using the restroom. Because of potential turbulence that could shake the plane, make sure you are holding on to the seat backs while navigating the aisle.

You may want to choose an aisle seat which will allow you to get up more easily to reach the restroom or just to stretch your legs and back. Although doubtful, the risk of deep vein thrombosis (DVT) can be further reduced by wearing compression stockings.

Risk factors that warrant travel considerations include the following:

- Severe anemia
- Cardiac disease
- Respiratory disease

- Recent hemorrhage
- Current or recent bone fractures

What precautions should I take when traveling by air?

- When you travel alone make sure you travel light and have the appropriate luggage.
- Make you sure you keep your cell phone with you at ALL times. This may be your only lifeline to call for help if need be.
- Travel at times when traffic is light to cut down travel time.
- On long trips, every hour, you should exercise your legs to prevent blood clots from forming.
- Stay hydrated by drinking non-alcoholic, non-caffeinated beverages.
- If possible, book direct flights. This allows you to avoid any layovers where oxygen may not be available.
- Check your health insurance coverage and travel insurance policy to make sure that any medical costs that may arise will be covered.
- Keep a current list of your medications while traveling.
- Bring the phone numbers of your health care providers, including your physician, respiratory therapist, and oxygen supplier.

"

Tough Times
Don't last

but

TOUGH PEOPLE

DO

"

13 | Pregnancy & Mental Health

"Mental health problems don't discriminate and neither should we- depression can be experienced by anyone, anytime... and depression during pregnancy is no exception."

Simone Honikman

More than one out of ten women of childbearing age struggle with depression, thus it is not uncommon to see depression as significant co-morbidity during pregnancy or immediately afterwards. An episode of depression is characterized by crying spells, guilty burden, appetite loss, lack of sleep, feelings of being sad, anxious, inability to concentrate, and having low energy. Untreated depression can increase risk of preterm labor, low birth weight, and decrease a baby's "Apgar score". The "Apgar score" is used to summarize the health of new born children. The scale is determined by evaluating the newborn baby on five simple criteria: appearance, pulse, grimace, activity, and respiration). However, effectively treating depression during pregnancy allows mothers to be nurturing to their body and developing baby.

Postpartum Depression

The birth of a baby is a time of delight and bliss, but it can also represent a time of great vulnerability of being mentally unwell, with postpartum depression (PPD) representing the most frequent form of maternal complication after childbirth. It can interfere with a mother's ability to care for herself or her child. It is an irritable and severe depression that can occur anytime during pregnancy or within 4 weeks of giving birth and up to a 1 year. The pattern of symptoms in mothers with postpartum depression is generally similar to those associated with major depression, apart from the fact that the content may focus on the delivery or baby.

Although, the cause of postpartum depression is not well understood, both physical/hormonal and emotional issues can play a significant role. While all women are susceptible to developing depression following childbirth, women who have one or more of the following factors are at a significantly higher risk.

- Previous history of postpartum depression
- History of depression prior to pregnancy
- Medical complications for mother or baby
- Pressure of Childcare
- Socioeconomic status
- Lack of support from partner, family or friends

Postpartum depression can interfere with mother-child bonding. Depressed mothers, are more likely to exhibit difficulties in engaging with their infant and are less likely to breast feed. These early disruptions can have intense impact on child development leading to behavioral problems and delayed cognition in the future.

The adverse impact of PPD upon the mother-child bond and child development makes the need for early identification and effective treatment a necessity. Therefore, it is pertinent to follow women with significant risk very closely after childbirth. Treatment for PPD should begin with psychotherapy (talk therapy) that can be combined with medication. Various types of psychotherapy techniques include interpersonal therapy (IBT) and short-term cognitive-behavioral therapy CBT) are available. If medications are used, then it should be given for sufficient duration of time to ensure complete recovery. Your physician can steer you towards the best medication during pregnancy for the treatment of depression.

The best thing a mother with postpartum depression can do is take care of herself, because while overcoming depression isn't quick or easy, it's far from impossible. In addition to talking with a professional, simple lifestyle changes can go a long way towards helping yourself feel better.

- Maintaining a healthy lifestyle that includes some physical activity, such as, taking a walk with your baby.

- Getting plenty of rest and good quality of sleep because poor sleep can make depression worse.
- Spending time talking with family and friends when you are feeling down. Talking with trusted people will help build an emotional connection that can guide you through tough time.
- Finding a support group. Joining a support group is an ideal way to connect with other mothers who also understand what you are going through. It is way to encourage each other and share experiences.

Only I can change
my life.

No one can do it
for me.

Appendix

List of Medications Used for Asthma

Bronchodilator Type	Generic Names	Trade Names®	Pregnancy Categories
Bronchodilator Nebulizer	Albuterol	Ventolin, Proventil	C
	Arformoterol	Brovana	C
	Levalbuterol	Xopenex	C
Bronchodilator HFA	Albuterol	Ventolin HFA, Proventil HFA, ProAir HFA	C
	Levalbuterol	Xopenex HFA	C
	Pirbutrol	Maxair	C
Bronchodilator (Anticholinergics)	Ipratropium Bromide Nebulizer/HFA	Atrovent Neb/HFA	B
	Tiotropium bromide	Spiriva*	B
	Aclidinium bromide	Tudorza	C
Bronchodilator (Anticholinergic + Beta Agonist)	Ipratropium Bromide with Albuterol	Combivent HFA*	C
Corticosteroids	Methylprednisolone	Medrol	C
	Prednisolone	Orapred	C
Inhaled Corticosteroids	Beclomethasone HFA	QVar	
	Budesonide Neb/DPI	Pulmicort Neb/DPI	B
	Ciclesonide HFA	Alvesco	C

	Flunisolide HFA	Aerospan	
	Fluticasone DPI/HFA	Flovent DPI/HFA	
	Mometasone	Asmanex	C
Combination Meds (Beta Agonist + Steroids)	Fluticasone/Salmeterol DPI/HFA	Advair DPI/HFA	C
	Mometasone/Formoterol HFA	Dulera HFA	C
	Budesonide/Formoterol HFA	Symbicort HFA	C
Phosphodiesterase 4 (PDE-4) Inhibitor	Roflumilast	Daliresp	C
Mast Cell Stabilizers	Cromolyn Sodium	Intal	B
Leukotriene Receptor Antagonist (LTRAs)	Montelukast	Singulair	B
	Zafirlukast	Accolate	B
	Zileuton	Zyflo	C
Monoclonal Antibody	Olamizumab	Xolair	B

*HFA – Hydrofluoroalkane, DPI – Dry Powdered Inhaler, * Also available with Respimat soft mist inhaler (SMI)

Table of Vaccinations

Vaccine	Before pregnancy	During pregnancy	After pregnancy	Type of Vaccine
Hepatitis A	Yes, if indicated	Yes, if indicated	Yes, if indicated	Inactivated
Hepatitis B	Yes, if indicated	Yes, if indicated	Yes, if indicated	Inactivated
Human Papillomavirus (HPV)	Yes, if indicated, through 26 years of age	No, under study	Yes, if indicated, through years of age	Inactivated
Influenza IIV	Yes	Yes	Yes	Inactivated
Influenza LAIV	Yes, if less than 50 years of age and healthy; avoid conception for 4 weeks	No	Yes, if less than 50 years of age and healthy; avoid conception for 4 weeks	Live
MMR	Yes, if indicated, avoid conception for 4 weeks	No	Yes, if indicated, give immediately postpartum if susceptible to rubella	Live
Meningococcal: • polysaccharide • conjugate	If indicated	If indicated	If indicated	Inactivated Inactivated
Pneumococcal Polysaccharide	If indicated	If indicated	If indicated	Inactivated

Tdap	Yes, if indicated	Yes, vaccinate during each pregnancy ideally between 27 and 36 weeks of gestation	Yes, immediately postpartum, if not received previously	Toxoid/inactivated
Tetanus/Diphtheria Td	Yes, if indicated	Yes, if indicated, Tdap preferred	Yes, if indicated	Toxoid
Varicella	Yes, if indicated, avoid conception for 4 weeks	No	Yes, if indicated, give immediately postpartum if susceptible	Live

Suggested Readings

1. Obstetrics and Gynecology 7th Edition. by Charles R. B. Beckmann MD MHPE, William Herbert MD, et al. ISBN: 978-1451144314

2. What to Expect When You're Expecting by Heidi Murkoff & Sharon Mazel. ISBN: 978-0761148579

3. Mayo Clinic Guide to a Healthy Pregnancy: From Doctors Who Are Parents, Too! by The Pregnancy Experts at Mayo Clinic. ISBN: 978-1561487172

4. March of Dimes Foundation - http://www.marchofdimes.org/

5. The American Congress of Obstetricians and Gynecologists - http://www.acog.org/Patients

6. American Pregnancy Association - http://americanpregnancy.org/

7. The American Heart Association - https://www.goredforwomen.org/

8. The Global Library of Women's Medicine - http://www.glowm.com/

9. e-Cigarettes. Sugerman DT. JAMA patient page. e-Cigarettes. JAMA. 2014 Jan 8;311(2):212.

www.ingramcontent.com/pod-product-compliance
Lightning Source LLC
Chambersburg PA
CBHW030854180526
45163CB00004B/1576